Modelling, Simulation and Control
of Non-Linear Dynamical Systems

Numerical Insights

Series Editor
A. Sydow, GMD-FIRST, Berlin, Germany

The Numerical Insights series aims to show how numerical simulations provide valuable insights into the mechanisms and processes involved in a wide range of disciplines. Such simulations provide a way of assessing theories by comparing simulations with observations. These models are also powerful tools which serve to indicate where both theory and experiment can be improved.

In most cases the books will be accompanied by software on disk demonstrating working examples of the simulations described in the text.

The editors will welcome proposals using modelling, simulation and systems analysis techniques in the following disciplines: physical sciences; engineering; environment; ecology; biosciences; economics.

Volume 1
Numerical Insights into Dynamic Systems: Interactive Dynamic System Simulation with Microsoft®, Windows 95™ and NT™
Granino A. Korn

Volume 2
Modelling, Simulation and Control of Non-Linear Dynamical Systems: An Intelligent Approach using Soft Computing and Fractal Theory
Patricia Melin and Oscar Castillo

Modelling, Simulation and Control of Non-Linear Dynamical Systems

An Intelligent Approach Using Soft Computing and Fractal Theory

Patricia Melin and Oscar Castillo

Tijuana Institute of Technology, Tijuana, Mexico

CRC Press
Taylor & Francis Group
Boca Raton London New York

CRC Press is an imprint of the
Taylor & Francis Group, an **informa** business
A TAYLOR & FRANCIS BOOK

First published 2002 by Taylor & francis

Published 2019 by CRC Press
Taylor & Francis Group
6000 Broken Sound Parkway NW, Suite 300
Boca Raton, FL 33487-2742

© 2002 by Taylor & Francis Group, LLC
CRC Press is an imprint of Taylor & Francis Group, an Informa business

First issued in paperback 2019

No claim to original U.S. Government works

ISBN 13: 978-0-367-45510-1 (pbk)
ISBN 13: 978-0-415-27236-0 (hbk)

**Visit the Taylor & Francis Web site at
http://www.taylorandfrancis.com**

**and the CRC Press Web site at
http://www.crcpress.com**

This book has been produced from camera-ready copy supplied by the authors

Every effort has been made to ensure that the advice and information in this book is true and accurate at the time of going to press. However, neither the publisher nor the authors can accept any legal responsibility or liability for any errors or omissions that may be made. In the case of drug administration, any medical procedure or the use of technical equipment mentioned within this book, you are strongly advised to consult the manufacturer's guidelines.

British Library Cataloguing in Publication Data
A catalogue record for this book is available from the British Library

Library of Congress Cataloging in Publication Data
A catalog record for this book has been requested

CONTENTS

PREFACE

This book presents a unified view of mathematical modelling, simulation and control for complex non-linear dynamical systems using soft computing techniques and fractal theory. Our particular point of view is that modelling, simulation and control are problems that can not be considered apart because they are intrinsically related in real-world applications. Control of non-linear dynamical systems can not be achieved if we don't have proper mathematical models for the systems. Also, useful simulations of a model, that can give us numerical insights into the behavior of a dynamical system, can not be obtained if we don't have the appropriate mathematical model. On the other hand, we have to recognize that complex non-linear dynamical systems can exhibit a wide range of dynamic behaviors (ranging from simple periodic orbits to chaotic strange attractors), so the problem of behavior identification is a very diffcult one. Also, we want to automate each of these tasks (modelling, simulation and control) because in this way it is easier to solve a particular problem. We then have three difficult tasks at hand: automated mathematical modelling of a dynamical system, automated simulation of the model, and model-based control of the system. A real world problem may require that we use modelling, simulation and control, to achieve the desired level of performance needed for the particular application.

Soft computing consists of several computing paradigms, including fuzzy logic, neural networks and genetic algorithms, which can be used to produce powerful hybrid intelligent systems. We believe that solving the difficult problems of modelling, simulation and control of non-linear dynamical systems require the use of several soft computing techniques to achieve the level of intelligence needed to automate the processes of modelling and simulation, and also to achieve adaptive control. On the other hand, fractal theory provides us with powerful mathematical tools that can be used to understand the geometrical complexity of natural or computational objects. We believe that, in many cases, it is necessary to use fractal tools to understand the geometry of the problem at hand. For example, the fractal dimension is a useful tool in measuring the geometrical complexity of a time series and for this reason can be used to formulate the corresponding mathematical model for the particular problem.

This book is intended to be a major reference for scientists and engineers interested in applying new computational and mathematical tools for solving the complicated problems of mathematical modelling, simulation and control of non-linear dynamical systems. The book can also be used at the graduate or advanced undergraduate level, as a textbook or major reference, for courses like: mathematical modelling, numerical simulation, non-linear control of dynamical systems, applied artificial intelligence and many others. We consider that this book can also be used to get new ideas for new lines of research or to continue the lines of future research proposed by the authors of the book. The software accompanying this book provides a good basis for developing more advanced 'intelligent' software tools for modelling, simulation and control of non-linear dynamical systems.

In Chapter 1, we begin by giving a brief introduction to the problems of modelling, simulation and control of non-linear dynamical systems. We motivate the importance of solving these problems, in an automated fashion, for real-world applications. We also outline the importance of using soft computing techniques and fractal theory to really achieve automated mathematical modelling and simulation, and model-based adaptive control of non-linear dynamical systems.

We present in Chapter 2 the main ideas underlying fuzzy logic and the application of this powerful computational theory to the problem of modelling. We discuss in some detail fuzzy set theory, fuzzy reasoning and fuzzy inference systems. At the end, we also give some remarks about fuzzy modelling. The importance of fuzzy logic as a basis for developing intelligent systems (sometimes in conjunction with other soft computing techniques) for control has been recognized in many areas of application. For this reason, we consider reading this chapter essential to understand the new methods for modelling, simulation and control presented in later chapters.

We present in Chapter 3 the basic concepts, notation and basic learning algorithms for neural networks. We discuss in some detail feedforward networks, adaptive neuro-fuzzy inference systems, neuro-fuzzy control and adaptive neuro-control. First, we give a brief review of the basic concepts of neural networks and the backpropagation learning algorithm. We then give a brief description of adaptive neuro-fuzzy systems. Finally, we end the chapter with a brief review on the current methods for neuro-fuzzy control and some remarks about adaptive control and model-based control. We can not emphasize enough the importance of neural networks as a computational tool to achieve 'intelligence' for software systems. For this reason, neural networks have been applied for solving complex problems of modelling, control and identification.

We present in Chapter 4 the basic concepts and notation of genetic algorithms, simulated annealing and fractal theory. Both genetic algorithms and simulated annealing are basic search methodologies that can be used for modelling and simulation of complex non-linear dynamical systems. Since both techniques can be considered as general purpose optimization methodologies, we can use them to find the mathematical model which minimizes the fitting errors for a specific problem. We also present in this chapter the basic concepts of dynamical systems and fractal theory, which are two powerful mathematical theories that enable the understanding of complex non-linear phenomena. Dynamical systems theory gives us the general framework for treating non-linear systems and enables the identification of the different dynamical behaviors that can occur for a particular dynamical system. On the other hand, fractal theory gives us powerful concepts and techniques that can be used to measure the complexity of geometrical objects.

We present in Chapter 5 our new method for automated mathematical modelling of non-linear dynamical systems. This method is based on a hybrid fuzzy-fractal approach to achieve, in an efficient way, automated modelling for a particular problem using a time series as a data set. The use of the fractal dimension is to perform time series analysis of the data, so as to obtain a qualitative characterization of the time series. The use of fuzzy logic techniques is to simulate the process of expert model selection using the qualitative information obtained from the time series analysis module. At the end, the 'best' mathematical model is obtained by comparing the measures of goodness for the selected mathematical models. In Chapter 8, we show some advanced applications of this method for automated mathematical modelling.

In Chapter 6, we describe the problem of numerical simulation for non-linear dynamical systems and its solution by using intelligent methodologies. The numerical simulation of a particular dynamical system consists in the successive application of a map and the subsequent identification of the corresponding dynamic behaviors. Automated simulation of a given dynamical system consists in selecting the appropriate parameter values for the model and then applying the corresponding iterative method (map) to find the limiting behavior. In this chapter, a new method for automated parameter selection, based on genetic algorithms, is introduced. Also, a new method for dynamic behavior identification, based on fuzzy logic, is introduced. The fuzzy-genetic approach for automated simulation consists in the integration of the method for automated parameter selection and the method for behavior identification.

We describe in Chapter 7 our new method for adaptive model-based control of non-linear dynamical systems. This method is based on a hybrid neuro-fuzzy approach to achieve, in an efficient way, adaptive robust control of non-linear dynamical systems using a set of different mathematical models. We use fuzzy logic to select the appropriate mathematical model for the dynamical system according to the changing conditions of the system. Adaptive control is achieved by using a neural network for control and a neural network for identification. Combining this method for control with the procedure for fuzzy model selection, gives us a new method for adaptive model-based control using a hybrid neuro-fuzzy approach. This method for adaptive control can be used for general dynamical systems or non-linear plants, since its architecture is domain independent. In Chapter 9, we show some advanced applications of this new method for adaptive model-based control.

In Chapter 8, we present several advanced applications of the new methods for automated mathematical modelling and simulation. First, we describe the application of the new methods for automated modelling and simulation to robotic dynamic systems, which is a very important application in the control of real-world robot arms and general robotic systems. Second, we apply our new methods for modelling and simulation to the problem of understanding the dynamic behavior of biochemical reactors in the food industry, which is also very important for the control of this type of dynamical system. Third, we consider the problem of modelling and simulation of international trade dynamics, which is an interesting problem in economics and finance. Finally, we also consider the problem of modelling and simulation of aircraft, as this is important for the real-world problem of automatic aircraft control.

In Chapter 9, we present several advanced applications of the new method for adaptive model-based control. First, we describe the application of the new method for adaptive model-based control to the case of robotic dynamic systems, which is very important for solving the problem of controlling real-world manipulators in real-time. Second, we describe the application of the method for adaptive model-based control to the case of biochemical reactors in the food industry, which is a very interesting case due to the complexity of this non-linear problem. Third, we consider briefly the problem of controlling international trade between three or more countries, with our new method for adaptive model-based control. Finally, we also consider briefly the problem of controlling aircraft with our new method for adaptive model-based control.

Finally, we would like to thank all the people who helped make this book possible. In particular, we would like to acknowledge our families for their love and support during the realization of this project; without them this book would never have been possible.

Chapter 1

Introduction to Modelling, Simulation and Control of Non-Linear Dynamical Systems

We describe in this book new methods for automated modelling and simulation of non-linear dynamical systems using Soft Computing techniques and Fractal Theory. We also describe a new method for adaptive model-based control of non-linear dynamical systems using a hybrid neuro-fuzzy-fractal approach. Soft Computing (SC) consists of several computing paradigms, including fuzzy logic, neural networks and genetic algorithms, which can be used to produce powerful hybrid intelligent systems. Fractal Theory (FT) provides us with the mathematical tools (like the fractal dimension) to understand the geometrical complexity of natural objects and can be used for identification and modelling purposes. Combining SC techniques with FT tools we can take advantage of the "intelligence" provided by the computer methods (like neural networks) and also take advantage of the descriptive power of fractal mathematical tools. Non-linear dynamical systems can exhibit extremely complex dynamic behavior and for this reason it is of great importance to develop intelligent computational tools that will enable the identification of the best model for a particular dynamical system, then obtaining the best simulations for the system and also achieving the goal of controlling the dynamical system in a desired manner. We also describe in this book the basic methodology to develop prototype intelligent systems that are able to find the best model for a particular dynamical system, then perform the numerical simulations necessary to identify all of the possible dynamical behaviors of the system, and finally achieve the goal of adaptive control using the mathematical models of the system and SC techniques.

As a prelude, we shall provide a brief overview of the existing methodologies for modelling, simulation and control of non-linear dynamical systems and also of our own approach in dealing with these problems.

1.1 Modelling and Simulation of Non-Linear Dynamical Systems

Traditionally, mathematical modelling of dynamical systems has been performed by human experts in the following manner (Jamshidi, 1997): 1) The expert according to his knowledge selects a set of models consider to be appropriate for a specific given problem, 2) Parameter estimation of the models is performed with methods similar to least-squares (using the relevant data available), and 3) The "best" model is selected using the measures of goodness for each of the models. Also, we can say that linear statistical models have been traditionally used as an approximation of real dynamic systems, which is not the best thing to do since many of the mechanical, electrical, biological and chemical systems are intrinsically non-linear in nature. In this work, we achieved automated mathematical modelling by using different Soft Computing techniques (Jang, Sun & Mizutani, 1997). The whole process of modelling starts with a time series (data set), which is used to perform a "Time Series Analysis" to extract the components of the time series (Weigend & Gershenfeld, 1994). Time series analysis can be achieved by traditional statistical methods or by efficient classification methods based on SC techniques, like neural networks or fuzzy logic (Kosko, 1997). In our case, we used fuzzy logic for classification of the time series components. After this time series analysis is performed, the qualitative values of the time series components are used to obtain a set of admissible models for a specific problem, this part of the problem was solved by using a set of fuzzy rules (knowledge base) that simulates the human experts in the domain of application. Finally, the "best" model is selected by comparing the measures of goodness for each of the admissible models considered in the previous step.

The simulation of mathematical models traditionally has been performed by exploring the possible dynamic behaviors, for a specific system, for different parameter values of the model (Rasband, 1990). More recently, it has been proposed to use Artificial Intelligence (Russell & Norvig, 1995) techniques for the simulation of mathematical models (for example, by using expert systems (Badiru, 1992)). In this work, we used SC techniques to automate the simulation of dynamical systems. In particular, we make use of genetic algorithms to generate the "best" set of parameter values for a specific model with respect to the goal of obtaining the most efficient simulation possible. Genetic Algorithms (GA) essentially consist of methods for the optimization of a general function based on the concept of "evolution" (Goldberg, 1989). In our particular case, the problem consisted in specifying the appropriate function to be optimized, with the goal of achieving the most efficient simulation possible, i.e., a simulation that enables the identification of all the possible dynamic behaviors for a specific dynamical system. For the identification of dynamic behaviors we make use of a fuzzy rule base that will identify a particular behavior according to the results of the numerical simulations.

In general, the study of non-linear dynamical systems is very important because most of the physical, electrical, mechanical and biochemical systems can be mathematically represented by models (differential or difference equations) in the time domain. Also, it is well known in Dynamical Systems Theory (Devaney, 1989) that the dynamic behavior of a particular system can range from very simple periodic orbits to the very complicated "chaotic" orbits. Non-linear models may exhibit the chaotic behavior for systems of at least three coupled differential equations or at least one difference equation (Ruelle, 1990). In particular, for the case of real-world dynamical systems the mathematical models needed are of very high dimensionality and in general there is a high probability of chaotic behavior, along with all sorts of different periodic and quasi-periodic behaviors (Castillo & Melin, 1998b). For this reason, it becomes very important to be able to obtain the appropriate mathematical models for the dynamical systems and then to be able to perform numerical simulations of these models (Castillo & Melin, 1997b), since this enables forecasting system's performance in future time. In this way, automated mathematical modelling and simulation of dynamical systems can contribute to real-time control of these systems, and this is critical in real-world applications (Melin & Castillo, 1998b). Also, an intelligent system for modelling and simulation can be useful in the design of real dynamical systems with certain constraints, since the information obtained by the numerical simulations can be used as a feedback in the process of design. The main contribution of the research work presented in this book is to combine several Soft Computing techniques to achieve automated mathematical modelling and simulation of non-linear dynamical systems using the advantages that each specific technique offers. For example, fuzzy logic (Von Altrock, 1995) was used to simulate the reasoning process of human experts in the process of mathematical modelling and genetic algorithms was used to select the best set of parameter values for the simulation of the best model.

The importance of the results presented in this book can be measured from the scientific point of view and also from the practical (or applications) point of view. First, from the scientific point of view, we consider that this research work is very important because the computer methods for automated mathematical modelling and simulation of dynamic systems that were developed contribute, in general, to the advancement of Computer Science, and, in particular, to the advancement of Soft Computing and Artificial Intelligence because the new algorithms that were developed can be considered "intelligent" in the sense that they simulate human experts in modelling and simulation. From the practical point of view, we consider the results of this research work very important for the areas of Control and Design of dynamical systems. Controlling dynamical systems can be made more easy if we are able to analyze and predict the dynamic evolution of these systems and this goal can be achieved with an intelligent system for automated mathematical modelling and simulation. The design of dynamical systems can be made more easy if we can use mathematical models and their simulations for planning the performance of these systems under different set of design constraints. This last two points are of great importance for the industrial applications, since the control of dynamical systems in real-world plants has to be very precise and also the design of this type of systems for specific tasks can be very useful for industry.

1.2 Control of Non-Linear Dynamical Systems

Traditional control of non-linear dynamical systems has been done by using Classical Linear Control Theory and assuming simple linear mathematical models for the systems. However, it is now well known that non-linear dynamical systems can exhibit complex behavior (and as a consequence are difficult to control) and the most appropriate mathematical models for them are the non-linear ones. Since the complexity of mathematical models for real dynamical systems is very high it becomes necessary to use more advanced control techniques. This is precisely the fact that motivated researchers in the area of Artificial Intelligence (AI) to apply techniques that mimic human experts in the domain of dynamical systems control. More recently, techniques like neural networks and fuzzy logic have been applied with some success to the control of non-linear dynamical systems for several domains of application. However, there also has been some limitations and problems with these approaches when applied to real systems. For this reason, we proposed in this book the application of a hybrid approach for the problem of control, combining neural and fuzzy technologies with the knowledge of the mathematical models for the adaptive control of dynamical systems. The basic idea of this hybrid approach is to combine the advantages of the computer methods with the advantages of using mathematical models for the dynamical systems. In this work, new methods were developed for adaptive control of non-linear systems using a combination of neural networks, fuzzy logic and mathematical models. Neural networks were used for the identification and control of the dynamical system and fuzzy logic was used to enable the change of mathematical models according to the dynamic state of the system. Also, the information and knowledge contained in the mathematical models was used for the control of the system by using their numerical results as input of the neural networks.

Traditionally, the control of dynamical systems has been performed using the classical methods of Linear Control Theory and also using linear models for the systems (Albertos, Strietzel & Mart, 1997). However, real-world problems can be viewed, in general, as non-linear dynamical systems with complex behavior and because of this, the most appropriate mathematical models for these systems are the non-linear ones. Unfortunately, to the moment, it hasn't been possible to generalize the results of Linear Control Theory to the case of Non-linear Control due to the complexity of the mathematics that will be required (Omidvar & Elliot, 1997). Of course, this mathematical generalization could still take several years of theoretical and empirical research to be developed. On the other hand, it is possible to use non-linear universal approximators that have resulted from the research in the area of SC to the problem of system identification and control. In particular, SC methodologies like neural networks and fuzzy logic have been applied with some success to problems of control and identification of dynamical systems (Korn, 1995). However, there are also problems where one or both methodologies have failed to achieved the level of accuracy desired in the applications (Omidvar & Elliot, 1997). For this reason, we have proposed in this work the use of a hybrid approach for the problem of non-linear adaptive control, i.e., we proposed to combine the use of neural networks and fuzzy logic with the use of non-linear mathematical models to achieve the goal of adaptive control. In the following lines we give the general idea of this new approach as well as the reasons why such an approach is a good alternative for non-linear control of dynamical systems.

Neural networks are computational systems with learning (or adaptive) characteristics that model the human brain (Kosko, 1992). Generally speaking, biological neural networks consist of neurons and connections between them and this is modeled by a graph with nodes and arcs to form the computational neural network. This graph along with a computational algorithm to specify the learning capabilities of the system is what makes the neural network a powerful methodology to simulate intelligent or expert behavior (Miller, Sutton & Werbos, 1995). It has been shown, that neural networks are universal approximators, in the sense that they can model any general function to a specified accuracy (Kosko, 1992) and for this reason neural networks have been applied to problems of system identification (Pham & Xing, 1995). Also, because of their adaptive capabilities neural networks have been used to control real-world dynamical systems (Ng, 1997).

Fuzzy Logic is an area of SC that enables a computer system to reason with uncertainty. Fuzzy inference systems consist of a set of "if-then" rules defined over fuzzy sets. Fuzzy sets are relations that can be used to model the linguistic variables that human experts use in their domain of expertise (Kosko, 1992). The main difference between fuzzy sets and traditional (crisp) sets is that the membership function for elements of a fuzzy set can take any value between 0 and 1, and not only 0 or 1. This corresponds, in the real world, to many situations where it is difficult to decide in an unambiguous manner if something belongs or not to a specific class. Fuzzy expert systems, for example, have been applied with some success to problems of control, diagnosis and classification just because they can manage the difficult expert reasoning involved in these areas of application (Korn, 1995). The main disadvantage with fuzzy systems is that they can't adapt to changing situations. For this reason, it is a good idea to combine both methodologies to have the advantages of neural networks (learning and adaptive capabilities) along with the advantages of fuzzy logic (contain expert knowledge) in solving complex real world problems where this flexibility is needed (Yen, Langar & Zadeh, 1995).

In this work, we have proposed a new architecture for developing intelligent control systems based on the use of neural networks, fuzzy logic and mathematical models, to achieve the goal of adaptive control of non-linear dynamical systems. The mathematical model of a non-linear dynamical system consist of a set of simultaneous non-linear differential (or difference) equations describing the dynamics of the system. The knowledge contained in the model is very important in the process of controlling the system, because it relates the different physical variables and their dependencies (Sueda & Iwamasa, 1995). For this reason, our approach is to combine mathematical models with neural networks and fuzzy logic, to achieve adaptive control of non-linear dynamical systems.

The study of non-linear dynamical systems is very interesting because of the complexity of the dynamics involved in the underlying processes (for example, biological, chemical or electrical) and also because of the implications, in the real world, of controlling industrial processes to maximize production. Real non-linear dynamical systems can have a wide range of possible dynamic behaviors, going from simple periodic orbits (stable) to the very complicated chaotic behavior (Kapitaniak, 1996). Controlling a non-linear dynamical system, avoiding chaotic behavior, is only possible using the mathematical models of the system (Sueda & Iwamasa, 1995). For this reason, model-based control is having great success in the control of complex real-world dynamical systems. In our approach, the neural networks were used for identification and control of the system, fuzzy logic was used to choose between different mathematical models of the system, and the knowledge given by the models was used to avoid specific and dangerous dynamic behaviors.

We consider the work on non-linear control presented in this book very important, from the point of view of Computer Science, because it contributed with new methods to develop intelligent control systems using a new hybrid model-based neuro-fuzzy approach for controlling non-linear dynamical systems. Also, from the point of view of the applications, this work is very important because it contributed with new methods for adaptive non-linear control that could eventually be used in the control of real industrial plants or general dynamical systems, which in turn will result in increased productivity and efficiency for these systems.

Chapter 2

Fuzzy Logic for Modelling

This chapter introduces the basic concepts, notation, and basic operations for fuzzy sets that will be needed in the following chapters. Since research on Fuzzy Set Theory has been underway for over 30 years now, it is practically impossible to cover all aspects of current developments in this area. Therefore, the main goal of this chapter is to provide an introduction to and a summary of the basic concepts and operations that are relevant to the study of fuzzy sets. We also introduce in this chapter the definition of linguistic variables and linguistic values and explain how to use them in fuzzy rules, which are an efficient tool for quantitative modelling of words or sentences in a natural or artificial language. By interpreting fuzzy rules as fuzzy relations, we describe different schemes of fuzzy reasoning, where inference procedures based on the concept of the compositional rule of inference are used to derive conclusions from a set of fuzzy rules and known facts. Fuzzy rules and fuzzy reasoning are the basic components of fuzzy inference systems, which are the most important modelling tool based on fuzzy set theory.

The "fuzzy inference system" is a popular computing framework based on the concepts of fuzzy set theory, fuzzy if-then rules, and fuzzy reasoning (Jang, Sun & Mizutani, 1997). It has found successful applications in a wide variety of fields, such as automatic control, data classification, decision analysis, expert systems, time series prediction, robotics, and pattern recognition (Jamshidi, 1997). Because of its multidisciplinary nature, the fuzzy inference system is known by numerous other names, such as "fuzzy expert system" (Kandel, 1992), "fuzzy model" (Sugeno & Kang, 1988), "fuzzy associative memory" (Kosko, 1992), and simply "fuzzy system".

The basic structure of a fuzzy inference system consists of three conceptual components: a "rule base", which contains a selection of fuzzy rules; a "data base" (or "dictionary"), which defines the membership functions used in the fuzzy rules; and a "reasoning mechanism", which performs the inference procedure upon the rules and given facts to derive a reasonable output or conclusion. In general, we can say that a fuzzy inference system implements a non-linear mapping from its input space to output space. This mapping is accomplished by a number of fuzzy if-then rules, each of which describes the local behavior of the mapping. In particular, the antecedent of a rule defines a fuzzy region in the input space, while the consequent specifies the output in the fuzzy region.

In what follows, we shall first introduce the basic concepts of fuzzy sets and fuzzy reasoning. Then we will introduce and compare the three types of fuzzy inference systems that have been employed in various applications. Finally, we will address briefly the features and problems of fuzzy modelling, which is concerned with the construction of fuzzy inference systems for modelling a given target system.

2.1 Fuzzy Set Theory

Let X be a space of objects and x be a generic element of X. A classical set A, $A \subseteq X$, is defined by a collection of elements or objects $x \in X$, such that each x can either belong or not belong to the set A. By defining a "characteristic function" for each element $x \in X$, we can represent a classical set A by a set of order pairs (x,0) or (x,1), which indicates $x \notin A$ or $x \in A$, respectively.

Unlike the aforementioned conventional set, a fuzzy set (Zadeh, 1965) expresses the degree to which an element belong to a set. Hence the characteristic function of a fuzzy set is allowed to have values between 0 and 1, which denotes the degree of membership of an element in a given set.

Definition 2.1 **Fuzzy sets and membership functions**
If X is a collection of objects denoted generically by x, then a "fuzzy set" A in X is defined as a set of ordered pairs:
$$A = \{ (x, \mu_A(x)) \mid x \in X \}. \tag{2.1}$$
where $\mu_A(x)$ is called "membership function" (or MF for short) for the fuzzy set A. The MF maps each element of X to a membership grade (or membership value) between 0 and 1.

Obviously, the definition of a fuzzy set is a simple extension of the definition of a classical set in which the characteristic function is permitted to have any values between 0 and 1. If the values of the membership function $\mu_A(x)$ is restricted to either 0 or 1, then A is reduced to a classical set and $\mu_A(x)$ is the characteristic function of A.

A fuzzy set is uniquely specified by its membership function. To describe membership functions more specifically, we shall define the nomenclature used in the literature (Jang, Sun & Mizutani, 1997).

Definition 2.2 **Support**
The "support" of a fuzzy set A is the set of all points x in X such that $\mu_A(x) > 0$:
$$\text{support} (A) = \{ x \mid \mu_A(x) > 0 \}. \tag{2.2}$$

Definition 2.3 **Core**
The "core" of a fuzzy set is the set of all points x in X such that $\mu_A(x) = 1$:
$$\text{core} (A) = \{ x \mid \mu_A(x) = 1 \}. \tag{2.3}$$

Definition 2.4 **Normality**
A fuzzy set A is "normal" if its core is nonempty. In other words, we can always find a point $x \in X$ such that $\mu_A(x) = 1$.

Definition 2.5 **Crossover points**

A "crossover point" of a fuzzy set A is a point $x \in X$ at which $\mu_A(x) = 0.5$:

$$\text{crossover}(A) = \{ x \mid \mu_A(x) = 0.5 \}. \tag{2.4}$$

Definition 2.6 **Fuzzy singleton**

A fuzzy set whose support is a single point in X with $\mu_A(x) = 1$ is called a "fuzzy singleton".

Corresponding to the ordinary set operations of union, intersection and complement, fuzzy sets have similar operations, which were initially defined in Zadeh's seminal paper (Zadeh, 1965). Before introducing these three fuzzy set operations, first we shall define the notion of containment, which plays a central role in both ordinary and fuzzy sets. This definition of containment is, of course, a natural extension of the case for ordinary sets.

Definition 2.7 **Containment**

The fuzzy set A is "contained" in fuzzy set B (or, equivalently, A is a "subset" of B) if and only if $\mu_A(x) \leq \mu_B(x)$ for all x. Mathematically,

$$A \subseteq B \Leftrightarrow \mu_A(x) \leq \mu_B(x). \tag{2.5}$$

Definition 2.8 **Union**

The "union" of two fuzzy sets A and B is a fuzzy set C, written as $C = A \cup B$ or C = A OR B, whose MF is related to those of A and B by

$$\mu_C(x) = \max(\mu_A(x), \mu_B(x)) = \mu_A(x) \vee \mu_B(x). \tag{2.6}$$

Definition 2.9 **Intersection**

The "intersection" of two fuzzy sets A and B is a fuzzy set C, written as $C = A \cap B$ or C = A AND B, whose MF is related to those of A and B by

$$\mu_C(x) = \min(\mu_A(x), \mu_B(x)) = \mu_A(x) \wedge \mu_B(x). \tag{2.7}$$

Definition 2.10 **Complement or Negation**

The "complement" of a fuzzy set A, denoted by $\overline{A}$ (⌐A, NOT A), is defined as

$$\mu_{\overline{A}}(x) = 1 - \mu_A(x). \tag{2.8}$$

As mentioned earlier, a fuzzy set is completely characterized by its MF. Since most fuzzy sets in use have a universe of discourse X consisting of the real line R, it would be impractical to list all the pairs defining a membership function. A more convenient and concise way to define an MF is to express it as a mathematical formula. First we define several classes of parameterized MFs of one dimension.

Definition 2.11 **Triangular MFs**

A "triangular MF" is specified by three parameters {a, b, c} as follows:

$$y = \text{triangle}(x;a,b,c) = \begin{cases} 0 & , & x \leq a . \\ (x\text{-}a) / (b\text{-}a) & , & a \leq x \leq b . \\ (c\text{-}x) / (c\text{-}b) & , & b \leq x \leq c . \\ 0 & , & c \leq x . \end{cases} \quad (2.9)$$

The parameters {a,b,c} (with a < b< c) determine the x coordinates of the three corners of the underlying triangular MF.

Figure 2.1 (a) illustrates a triangular MF defined by triangle(x; 10, 20, 40).

Definition 2.12 **Trapezoidal MFs**

A "trapezoidal MF" is specified by four parameters {a, b, c, d} as follows:

$$\text{trapezoid }(x;a,b,c,d) = \begin{cases} 0 & , & x \leq a . \\ (x\text{-}a) / (b\text{-}a) & , & a \leq x \leq b . \\ 1 & , & b \leq x \leq c . \\ (d\text{-}x) / (d\text{-}c) & , & c \leq x \leq d . \\ 0 & , & d \leq x . \end{cases} \quad (2.10)$$

The parameters {a, b, c, d} (with a < b ≤ c <d) determine the x coordinates of the four corners of the underlying trapezoidal MF.

Figure 2.1 (b) illustrates a trapezoidal MF defined by trapezoid(x; 10, 20 40, 75).

Due to their simple formulas and computational efficiency, both triangular MFs and trapezoidal MFs have been used extensively, especially in real-time implementations. However, since the MFs are composed of straight line segments, they are not smooth at the corner points specified by the parameters. In the following we introduce other types of MFs defined by smooth and nonlinear functions.

<u>Definition</u> 2.13 **Gaussian MFs**

A "Gaussian MF" is specified by two parameters $\{c, \sigma\}$:

$$\text{gaussian}(x; c, \sigma) = e^{-\frac{1}{2}\left(\frac{x-c}{\sigma}\right)^2}. \tag{2.11}$$

A "Gaussian MF is determined completely by c and σ; c represents the MFs center and σ determines the MFs width. Figure 2.2 (a) plots a Gaussian MF defined by gaussian (x; 50, 20).

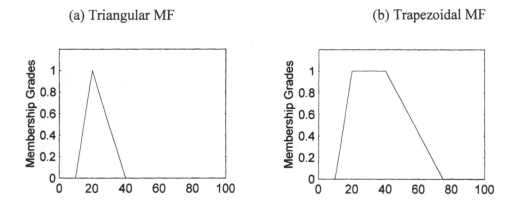

(a) Triangular MF   (b) Trapezoidal MF

Figure 2.1 Examples of two types of parameterized MFs

<u>Definition</u> 2.14 **Generalized bell MFs**

A "generalized bell MF" is specified by three parameters $\{a, b, c\}$:

$$\text{bell}(x; a, b, c) = \frac{1}{1 + |(x-c)/a|^{2b}} \tag{2.12}$$

where the parameter b is usually positive. We can note that this MF is a direct generalization of the Cauchy distribution used in probability theory, so it is also referred to as the "Cauchy MF".

Figure 2.2 (b) illustrates a generalized bell MF defined by bell(x; 20, 4, 50).

Although the Gaussian MFs and bell MFs achieve smoothness, they are unable to specify asymmetric MFs, which are important in certain applications. Next we define the sigmoidal MF, which is either open left or right.

(a) Gaussian MF   (b) Generalized Bell MF

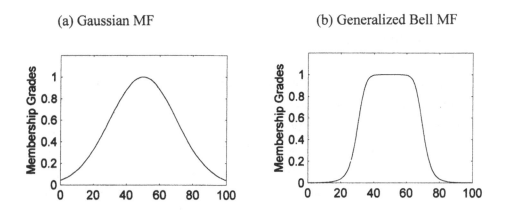

Figure 2.2 Examples of two classes of parameterized continuous MFs.

Definition 2.15 **Sigmoidal MFs**

A "Sigmoidal MF" is defined by the following equation:

$$sig(x; a, c) = \frac{1}{1 + \exp[-a(x-c)]} \qquad (2.13)$$

where a controls the slope at the crossover point $x = c$.

Depending on the sign of the parameter "a", a sigmoidal MF is inherently open right or left and thus is appropriate for representing concepts such as "very large" or "very negative". Figure 2.3 shows two sigmoidal functions $y_1 = sig(x; 1, -5)$ and $y_2 = sig(x; -2, 5)$.

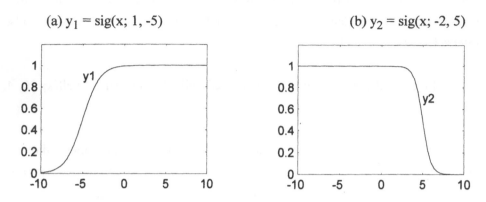

Figure 2.3 Two sigmoidal functions y_1 and y_2 .

2.2 Fuzzy Reasoning

As was pointed out by Zadeh in his work on this area (Zadeh, 1973), conventional techniques for system analysis are intrinsically unsuited for dealing with humanistic systems, whose behavior is strongly influenced by human judgment, perception, and emotions. This is a manifestation of what might be called the "principle of incompatibility": "As the complexity of a system increases, our ability to make precise and yet significant statements about its behavior diminishes until a threshold is reached beyond which precision and significance become almost mutually exclusive characteristics" (Zadeh, 1973). It was because of this belief that Zadeh proposed the concept of linguistic variables (Zadeh, 1971) as an alternative approach to modelling human thinking.

Definition 2.16 **Linguistic variables**
A "Linguistic variable" is characterized by a quintuple (x, T(x), X, G, M) in which x is the name of the variable; T(x) is the "term set" of x-that is, the set of its "linguistic values" or "linguistic terms"; X is the universe of discourse, G is a "syntactic rule" which generates the terms in T(x); and M is a "semantic rule" which associates with each linguistic value A its meaning M(A), where M(A) denotes a fuzzy set in X.

<u>Definition</u> 2.17 **Concentration and dilation of linguistic values**

Let A be a linguistic value characterized by a fuzzy set membership function $\mu_A(.)$. Then A^k is interpreted as a modified version of the original linguistic value expressed as

$$A^k = \int_x [\, \mu_A(x)]^k / x \; . \tag{2.14}$$

In particular, the operation of "concentration" is defined as

$$CON(A) = A^2 \; , \tag{2.15}$$

while that of "dilatation" is expressed by

$$DIL(A) = A^{0.5} \; . \tag{2.16}$$

Conventionally, we take CON(A) and DIL(A) to be the results of applying the hedges "very" and "more or less", respectively, to the linguistic term A. However, other consistent definitions for these linguistic hedges are possible and well justified for various applications.

Following the definitions given before, we can interpret the negation operator NOT and the connectives AND and OR as

$$NOT(A) = \rceil A = \int_x [\, 1 - \mu_A(x)\,] / x \;\; ,$$
$$A \; AND \; B = A \cap B = \int_x [\, \mu_A(x) \wedge \mu_B(x)\,] / x \;\; , \tag{2.17}$$
$$A \; OR \; B = A \cup B = \int_x [\, \mu_A(x) \vee \mu_B(x)\,] / x$$

respectively, where A and B are two linguistic values whose meanings are defined by $\mu_A(.)$ and $\mu_B(.)$.

<u>Definition</u> 2.18 **Fuzzy If-Then Rules**

A "fuzzy if-then rule" (also known as "fuzzy rule", "fuzzy implication", or "fuzzy conditional statement") assumes the form

$$\text{if } x \text{ is } A \text{ then } y \text{ is } B \;\; , \tag{2.18}$$

where A and B are linguistic values defined by fuzzy sets on universes of discourse X and Y, respectively. Often "x is A" is called "antecedent" or "premise", while "y is B" is called the "consequence" or "conclusion".

Examples of fuzzy if-then rules are widespread in our daily linguistic expressions, such as the following:

- If pressure is high, then volume is small.
- If the road is slippery, then driving is dangerous.
- If the speed is high, then apply the brake a little.

Before we can employ fuzzy if-then rules to model and analyze a system, first we have to formalize what is meant by the expression "if x is A then y is B", which is sometimes abbreviated as $A \rightarrow B$. In essence, the expression describes a relation between two variables x and y; this suggests that a fuzzy if-then rule is defined as a binary fuzzy relation R on the product space $X \times Y$. Generally speaking, there are two ways to interpret the fuzzy rule $A \rightarrow B$. If we interpret $A \rightarrow B$ as A "coupled with" B then

$$R = A \rightarrow B = A \times B = \int_{X \times Y} \mu_A(x) \,\widetilde{*}\, \mu_B(y) \,/\, (x,y)$$

where $\widetilde{*}$ is an operator for intersection (Mamdani & Assilian, 1975). On the other hand, if $A \rightarrow B$ is interpreted as A "entails" B, then it can be written as one of two different formulas:

- Material implication:
$$R = A \rightarrow B = \neg A \cup B . \tag{2.19}$$
- Propositional Calculus:
$$R = A \rightarrow B = \neg A \cup (A \cap B) . \tag{2.20}$$

Although these two formulas are different in appearance, they both reduce to the familiar identity $A \rightarrow B \equiv \neg A \cup B$ when A and B are propositions in the sense of two-valued logic.

Fuzzy reasoning, also known as approximate reasoning, is an inference procedure that derives conclusions from a set of fuzzy if-then rules and known facts. The basic rule of inference in traditional two-valued logic is "modus ponens", according to which we can infer the truth of a proposition B from the truth of A and the implication A → B. This concept is illustrated as follows:

premise 1 (fact):	x is A ,
premise 2 (rule):	if x is A then y is B ,
consequence (conclusion):	y is B .

However, in much of human reasoning, modus ponens is employed in an approximate manner. This is written as

premise 1 (fact):	x is A'
premise 2 (rule):	if x is A then y is B ,
consequence (conclusion):	y is B'

where A' is close to A and B' is close to B. When A, B, A' and B' are fuzzy sets of appropriate universes, the foregoing inference procedure is called "approximate reasoning" or "fuzzy reasoning"; it is also called "generalized modus ponens" (GMP for short), since it has modus ponens as a special case.

Definition 2.19 **Fuzzy reasoning**

Let A, A', and B be fuzzy sets of X, X, and Y respectively. Assume that the fuzzy implication A → B is expressed as a fuzzy relation R on X × Y .Then the fuzzy set B induced by "x is A'" and the fuzzy rule "if x is A then y is B" is defined by

$$\mu_{B'}(y) = \max_x \min [\mu_{A'}(x), \mu_R(x, y)]$$
$$= V_x [\mu_{A'}(x) \wedge \mu_R(x, y)] . \qquad (2.21)$$

Now we can use the inference procedure of fuzzy reasoning to derive conclusions provided that the fuzzy implication A → B is defined as an appropriate binary fuzzy relation.

Single Rule with Single Antecedent

This is the simplest case, and the formula is available in Equation (2.21). A further simplification of the equation yields

$$\mu_{B'}(y) = [V_x (\mu_{A'}(x) \wedge \mu_A(x))] \wedge \mu_B(y)$$
$$= \omega \wedge \mu_B(y)$$

In other words, first we find the degree of match ω as the maximum of $\mu_{A'}(x) \wedge \mu_A(x)$; then the MF of the resulting B' is equal to the MF of B clipped by ω. Intuitively, ω represents a measure of degree of belief for the antecedent part of a rule; this measure gets propagated by the if-then rules and the resulting degree of belief or MF for the consequent part should be no greater than ω.

Multiple Rules with Multiple Antecedents

The process of fuzzy reasoning or approximate reasoning for the general case can be divided into four steps:

1) Degrees of compatibility: Compare the known facts with the antecedents of fuzzy rules to find the degrees of compatibility with respect to each antecedent MF.
2) Firing Strength: Combine degrees of compatibility with respect to antecedent MFs in a rule using fuzzy AND or OR operators to form a firing strength that indicates the degree to which the antecedent part of the rule is satisfied.
3) Qualified (induced) consequent MFs: Apply the firing strength to the consequent MF of a rule to generate a qualified consequent MF.
4) Overall output MF: Aggregate all the qualified consequent MFs to obtain an overall output MF.

2.3 Fuzzy Inference Systems

The "Mamdani fuzzy inference system" (Mamdani & Assilian, 1975) was proposed as the first attempt to control a steam engine and boiler combination by a set of linguistic control rules obtained from experienced human operators. Figure 2.4 is an illustration of how a two-rule Mamdani fuzzy inference system derives the overall output z when subjected to two numeric inputs x and y.

In Mamdani's application, two fuzzy inference systems were used as two controllers to generate the heat input to the boiler and throttle opening of the engine cylinder, respectively, to regulate the steam pressure in the boiler and the speed of the engine. Since the engine and boiler take only numeric values as inputs, a defuzzifier was used to convert a fuzzy set to a numeric value.

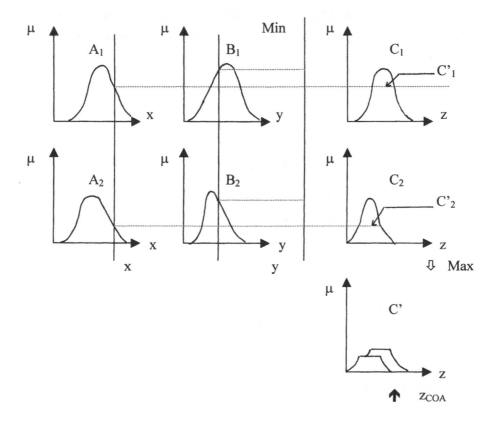

Figure 2.4 The Mamdani fuzzy inference system using the min and max operators.

Defuzzification

Defuzzification refers to the way a numeric value is extracted from a fuzzy set as a representative value. In general, there are five methods for defuzzifying a fuzzy set A of a universe of discourse Z, as shown in Figure 2.5 (Here the fuzzy set A is usually represented by an aggregated output MF, such as C' in Figure 2.4). A brief explanation of each defuzzification strategy follows.

- Centroid of area z_{COA}:

$$z_{COA} = \frac{\int_z \mu_A(z)z dz}{\int_z \mu_A(z)dz} \qquad (2.22)$$

where $\mu_A(z)$ is the aggregated output MF. This is the most widely adopted defuzzification strategy, which is reminiscent of the calculation of expected values of probabillty distributions.

- Bisector of area z_{BOA} : z_{BOA} satisfies

$$\int_\alpha^{z_{BOA}} \mu_A(z)dz = \int_{z_{BOA}}^\beta \mu_A(z)dz \quad, \qquad (2.23)$$

where $\alpha = \min\{z \mid z \in Z\}$ and $\beta = \max\{z \mid z \in Z\}$.

- Mean of maximum z_{MOM} : z_{MOM} is the average of the maximizing z at which the MF reach a maximum μ^*. Mathematically,

$$z_{MOM} = \frac{\int_{z'} z dz}{\int_{z'} dz} \quad, \qquad (2.24)$$

where $z' = \{ z \mid \mu_A(z) = \mu^* \}$. In particular, if $\mu_A(z)$ has a single maximum at $z = z^*$, then $z_{MOM} = z^*$. Moreover, if $\mu_A(z)$ reaches its maximum whenever $z \in [z_{left}, z_{right}]$ then $z_{MOM} = (z_{left} + z_{right}) / 2$.

- Smallest of maximum z_{SOM} : z_{SOM} is the minimum (in terms of magnitude) of the maximizing z.

- Largest of maximum z_{LOM} : z_{LOM} is the maximum (in terms of magnitude) of the maximizing z. Because of their obvious bias, z_{SOM} and z_{LOM} are not used as often as the other three defuzzification methods.

The calculation needed to carry out any of these five defuzzification operations is time-consuming unless special hardware support is available. Furthermore, these defuzzification operations are not easily subject to rigorous mathematical analysis, so most of the studies are based on experimental results. This leads to the propositions of other types of fuzzy inference systems that do not need defuzzification at all; two of them will be described in the following. Other more flexible defuzzification methods can be found in several more recent papers (Yager & Filer, 1993), (Runkler & Glesner, 1994).

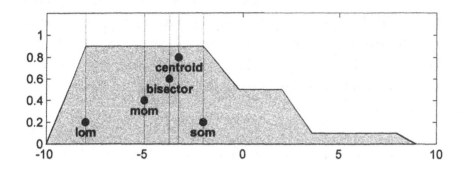

Figure 2.5 Various defuzzification methods for obtaining a numeric output.

Sugeno Fuzzy Models

The "Sugeno fuzzy model" (also known as the "TSK fuzzy model") was proposed by Takagi, Sugeno and Kang in an effort to develop a systematic approach to generating fuzzy rules from a given input-output data set (Takagi & Sugeno, 1985), (Sugeno & Kang, 1988). A typical fuzzy rule in a sugeno fuzzy model has the form:

if x is A and y is B then z = f(x,y)

where A and B are fuzzy sets in the antecedent, while $z = f(x,y)$ is a traditional function in the consequent. Usually $f(x,y)$ is a polynomial in the input variables x and y, but it can be any function as long as it can appropriately describe the output of the model within the fuzzy region specified by the antecedent of the rule. When $f(x,y)$ is a first-order polynomial, the resulting fuzzy inference system is called a "first-order Sugeno fuzzy model". When f is constant, we then have a "zero-order Sugeno fuzzy model", which can be viewed either as a special case of the Mamdani inference system, in which each rule's consequent is specified by a fuzzy singleton, or a special case of the Tsukamoto fuzzy model (to be introduced next), in which each rule's consequent is specified by an MF of a step function center at the constant.

Figure 2.6 shows the fuzzy reasoning procedure for a first-order Sugeno model. Since each rule has a numeric output, the overall output is obtained via "weighted average", thus avoiding the time-consuming process of defuzzification required in a Mamdani model. In practice, the weighted average operator is sometimes replaced with the "weighted sum" operator (that is, $w_1z_1 + w_2z_2$ in Figure 2.6) to reduce computation further, specially in the training of a fuzzy inference system. However, this simplification could lead to the loss of MF linguistic meanings unless the sum of firing strengths (that is, $\sum w_i$) is close to unity.

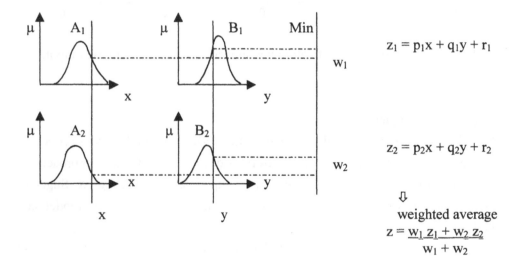

$$z_1 = p_1x + q_1y + r_1$$

$$z_2 = p_2x + q_2y + r_2$$

$$\Downarrow$$

weighted average
$$z = \frac{w_1 z_1 + w_2 z_2}{w_1 + w_2}$$

Figure 2.6 The Sugeno fuzzy model.

Tsukamoto Fuzzy Models

In the "Tsukamoto fuzzy models" (Tsukamoto, 1979), the consequent of each fuzzy if-then rule is represented by a fuzzy set with a monotonical MF, as shown in Figure 2.7. As a result, the inferred output of each rule is defined as a numeric value induced by the rule firing strength. The overall output is taken as the weighted average of each rule's output. Figure 2.7 illustrates the reasoning procedure for a two-input two-rule system.

Since each rule infers a numeric output, the Tsukamoto fuzzy model aggregates each rule's output by the method of weighted average and thus avoids the time-consuming process of defuzzification. However, the Tsukamoto fuzzy model is not used often since it is not as transparent as either the Mamdani or Sugeno fuzzy models.

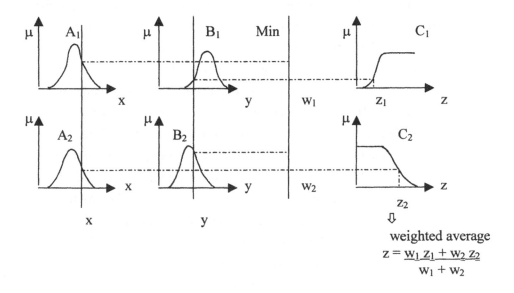

Figure 2.7 The Tsukamoto fuzzy model.

There are certain common issues concerning all the three fuzzy inference systems introduced previously, such as how to partition an input space and how to construct a fuzzy inference system for a particular application.

Input Space Partitioning

Now it should be clear that the spirit of fuzzy inference systems resembles that of "divide and conquer" - the antecedent of a fuzzy rule defines a local fuzzy region, while the consequent describes the behavior within the region via various constituents. The consequent constituent can be a consequent MF (Mamdani and Tsukamoto fuzzy models), a constant value (zero-order Sugeno model), or a linear equation (first-order Sugeno model). Different consequent constituents result in different fuzzy inference systems, but their antecedents are always the same. Therefore, the following discussion of methods of partitioning input spaces to form the antecedents of fuzzy rules is applicable to all three types of fuzzy inference systems.

- Grid partition: This partition method is often chosen in designing a fuzzy controller, which usually involves only several state variables as the inputs to the controller. This partition strategy needs only a small number of MFs for each input. However, it encounters problems when we have a moderately large number of inputs. For instance, a fuzzy model with 10 inputs and 2 MFs on each input would result in $2^{10} = 1024$ fuzzy if-then rules, which is prohibitively large. This problem, usually referred to as the "curse of dimensionality", can be alleviated by other partition strategies.

- Tree partition: In this method each region can be uniquely specified along a corresponding decision tree. The tree partition relieves the problem of an exponential increase in the number of rules. However, more MFs for each input are needed to define these fuzzy regions, and these MFs do not usually bear clear linguistic meanings. In other words, ortogonality holds roughly in $X \times Y$, but not in either X or Y alone.

- Scatter partition: By covering a subset of the whole input space that characterizes a region of possible occurrence of the input vectors, the scatter partition can also limit the number of rules to a reasonable amount. However, the scatter partition is usually dictated by desired input-output data pairs and thus, in general, orthogonality does not hold in X, Y or $X \times Y$. This makes it hard to estimate the overall mapping directly from the consequent of each rule's output.

2.4 Fuzzy Modelling

In general, we design a fuzzy inference system based on the past known behavior of a target system. The fuzzy system is then expected to be able to reproduce the behavior of the target system. For example, if the target system is a human operator in charge of a chemical reaction process, then the fuzzy inference system becomes a fuzzy logic controller that can regulate and control the process.

Let us now consider how we might construct a fuzzy inference system for a specific application. Generally speaking, the standard method for constructing a fuzzy inference system, a process usually called "fuzzy modelling", has the following features:

- The rule structure of a fuzzy inference system makes it easy to incorporate human expertise about the target system directly into the modelling process. Namely, fuzzy modelling takes advantage of "domain knowledge" that might not be easily or directly employed in other modelling approaches.
- When the input-output data of a target system is available, conventional system identification techniques can be used for fuzzy modelling. In other words, the use of "numerical data" also plays an important role in "fuzzy modelling", just as in other mathematical modelling methods.

Conceptually, fuzzy modelling can be pursued in two stages, which are not totally disjoint. The first stage is the identification of the "surface structure", which includes the following tasks:

1. Select relevant input and output variables.
2. Choose a specific type of fuzzy inference system.
3. Determine the number of linguistic terms associated with each input and output variables.
4. Design a collection of fuzzy if-then rules.

Note that to accomplish the preceding tasks, we rely on our own knowledge (common sense, simple physical laws, an so on) of the target system, information provided by human experts who are familiar with the target system, or simply trial and error.

After the first stage of fuzzy modelling, we obtain a rule base that can more or less describe the behavior of the target system by means of linguistic terms. The meaning of these linguistic terms is determined in the second stage, the identification of "deep structure", which determines the MFs of each linguistic term. Specifically, the identification of deep structure includes the following tasks:

1. Choose an appropriate family of parameterized MFs.
2. Interview human experts familiar with the target systems to determine the parameters of the MFs used in the rule base.
3. Refine the parameters of the MFs using regression and optimization techniques.

Task 1 and 2 assume the availability of human experts, while task 3 assumes the availability of a desired input-output data set.

2.5 Summary

In this chapter, we have presented the main ideas underlying Fuzzy Logic and we have only started to point out the many possible applications of this powerful computational theory. We have discussed in some detail fuzzy set theory, fuzzy reasoning and fuzzy inference systems. At the end, we also gave some remarks about fuzzy modelling. In the following chapters, we will show how fuzzy logic techniques (in some cases, in conjunction with other methodologies) can be applied to solve real world complex problems. This chapter will serve as a basis for the new hybrid intelligent methods, for modelling and simulation, that will be described in Chapters 5 and 6 of this book. Fuzzy Logic will also play an important role in the new neuro-fuzzy methodology for control that is presented in Chapter 7 of this book.

Chapter 3

Neural Networks for Control

Application of fuzzy inference systems to automatic control was first reported in Mamdani's paper (Mamdani & Assilian, 1975), where a "fuzzy logic controller" (FLC) was used to emulate a human operator's control of a steam engine and boiler combination. Since then, "fuzzy logic control" has been recognized as the most significant and fruitful application for fuzzy logic (Kosko, 1992). In the past few years, advances in microprocessors and hardware technologies have created an even more diversified application domain for fuzzy logic controllers, which ranges from consumer electronics to the automobile industry. However, without adaptive capability, the performance of FLCs relies exclusively on two factors: the availability of human experts, and the knowledge acquisition techniques to convert human expertise into appropriate fuzzy rules. These two factors substantially restrict the application domain of FLCs.

On the other hand, investigation into using neural networks in automatic control systems did not receive much attention until the "backpropagation" learning rule was formulated by Rumelhart and others (Rumelhart, Hinton & Williams, 1986). Since then, research of neural control has evolved quickly and a number of neural controller design methods have been proposed in the literature (Werbos, 1991).

Figure 3.1 is a block diagram of a typical "feedback control system", where the "plant" (or "process") represents the dynamic system to be controlled and the "controller" employs a control strategy to achieve a control goal. Here we shall denote the state variables of the plant as a vector x(t); these variables are usually governed by a set of "state equations" (usually differential equations) that characterize the dynamic behavior of the plant. Since the state variables are internal to the plant, some of them may not be directly measurable from the external world. The measurable quantities of the plant, also known as its outputs, are denoted as a vector y(t). We shall assume that all states are measurable; thus the output of the plant y(t) is equal to the state x(t).

The state equation for a general non-linear plant can be expressed in the matrix notation

$$\mathbf{x}'(t) = \mathbf{f}\,(\mathbf{x}(t),\, \mathbf{u}(t)) \qquad \text{(plant dynamics)} \qquad (3.1)$$

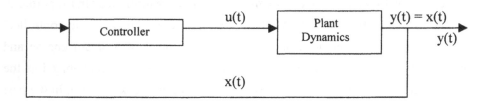

Figure 3.1 Block diagram for a continuous feedback control system.

where u(t) is the controller's output at time t, and the size of the vector $\mathbf{x}(t)$ is called the "order" of the plant. A general control goal is to find a controller with a static function ϕ that maps an observed plant output $\mathbf{x}(t)$ to a control action $\mathbf{u}$-that is, $\mathbf{u}(t) = \phi\,(\mathbf{x}(t))$- such that the plant output can follow some given desired output signal $\mathbf{x}_d(t)$ as closely as possible. If $\mathbf{x}_d(t)$ is a constant vector, then the control problem is referred to as "regulator problem", where the plant states are directly fed back to the controller. This is actually what Figure 3.1 shows. On the other hand, if the desired trajectory $\mathbf{x}_d(t)$ is a time-varying signal, then we have a "tracking problem" in which an error signal, defined as the difference between desired and actual outputs, is fed back to the controller. If **f** is unknown, we need to perform system identification first to find a model for the plant. Moreover, if **f** is time varying, it is desirable to make ϕ adaptive to respond to the changing characteristics of the plant.

In the case of linear feedback control systems, the plant and controller can be reformulated as the following equations:

$$\mathbf{x}'(t) = \mathbf{A}\mathbf{x}(t) + \mathbf{B}\mathbf{u}(t) \qquad \text{(plant dynamics)} \qquad (3.2)$$
$$\mathbf{u}(t) = \mathbf{k}\mathbf{x}(t) \qquad \text{(linear controller)}$$

The treatment of linear control systems is relatively complete in the literature (for example, see Brogan, 1991) and will not be discussed here. On the other hand, the area of non-linear control is still with many open problems and its more interesting. In this book, the treatment will be restricted to non-linear plants with a general form given by Equation (3.1).

If we replace the controller block in Figure 3.1 with neural networks or fuzzy systems, then we end up with "neural" or "fuzzy control systems", respectively. In other words, neural or fuzzy control design methods are systematic ways of constructing neural networks or fuzzy inference systems, respectively, as controllers intended to achieve prescribed control goals. In the same vein, the term "neuro-fuzzy control" has been used when one is speaking about design methods for fuzzy logic controllers that use neural network techniques.

Most neural or fuzzy controllers are nonlinear; thus rigorous analysis for neuro-fuzzy control systems is difficult and remains a challenging area for further investigation. On the other hand, a neuro-fuzzy controller usually contains a large number of parameters; it is thus more versatile than a linear controller in dealing with non-linear plant characteristics. Therefore, neuro-fuzzy controllers almost always surpass pure linear controllers if designed properly.

In this chapter, we present the basic concepts, notation, and basic learning algorithms for neural networks that will be needed in the following chapters of this book. The chapter is organized as follows: Backpropagation for Feedforward Networks, Adaptive Neuro-Fuzzy Inference Systems, Neuro-Fuzzy Control and

Adaptive Neuro-Control. First, we give a brief review of the basic concepts of neural networks and the backpropagation learning algorithm. Second, we give a brief description of adaptive neuro-fuzzy systems. Third, we give a brief review on the current methods for neuro-fuzzy control. Finally, we end the chapter with some remarks about adaptive control and model-based control. We consider this material necessary to understand the new methods for control that will be presented in Chapter 7 of this book.

3.1 Backpropagation for Feedforward Networks

This section describes the architectures and learning algorithms for adaptive networks, a unifying framework that subsumes almost all kinds of neural network paradigms with supervised learning capabilities. An adaptive network, as the name indicates, is a network structure consisting of a number of nodes connected through directional links. Each node represents a process unit, and the links between nodes specify the causal relationship between the connected nodes. The learning rule specifies how the parameters (of the nodes) should be updated to minimize a prescribed error measure.

The basic learning rule of the adaptive network is the well-known steepest descent method, in which the gradient vector is derived by successive invocations of the chain rule. This method for systematic calculation of the gradient vector was proposed independently several times, by Bryson and Ho (1969), Werbos (1974), and Parker (1982). However, because research on artificial neural networks was still in its infancy at those times, these researchers' early work failed to receive the attention it deserved. In 1986, Rumelhart et al. used the same procedure to find the gradient in a multilayer neural network. Their procedure was called "backpropagation learning rule", a name which is now widely known because the work of Rumelhart inspired enormous interest in research on neural networks. In this section, we introduce Werbos's original backpropagation method for finding gradient vectors and also present improved versions of this method.

3.1.1 The Backpropagation learning algorithm

Suppose that a given feedforward adaptive network has L layers and layer l (l = 0, 1,..., L) has N(l) nodes. Then the output and function of node i [i = 1, ..., N(l)] in layer l can be represented as $x_{l,i}$ and $f_{l,i}$, respectively, as shown in Figure 3.2. Since the output of a node depends on the incoming signals and the parameter set of the node, we have the following general expression for the node function $f_{l,i}$:

$$x_{l,i} = f_{l,i} (x_{l-1,1} , ... , x_{l-1,N(l-1)}, \alpha, \beta, \gamma , ...) \qquad (3.3)$$

where α, β, γ , etc. are the parameters of this node.

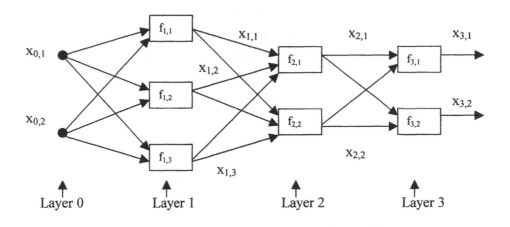

Figure 3.2 Feedforward adaptive network.

Assuming that the given training data set has P entries, we can define an error measure for the pth (1 ≤ p ≤ P) entry of the training data as the sum of the squared errors:

$$E_p = \sum_{k=1}^{N(L)} (d_k - x_{L,k})^2 \qquad (3.4)$$

where d_k is the kth component of the pth desired output vector and $x_{L,k}$ is the kth component of the actual output vector produced by presenting the pth input vector to the network. Obviously, when E_p is equal to zero, the network is able to reproduce exactly the desired output vector in the pth training data pair. Thus our task here is to minimize an overall error measure, which is defined as $E = \sum E_p$.

We can also define the "error signal" $\varepsilon_{l,i}$ as the derivative of the error measure E_p with respect to the output of the node i in layer l, taking both direct and indirect paths into consideration. Mathematically,

$$\varepsilon_{l,i} = \frac{\partial^+ E_p}{\partial x_{l,i}} \tag{3.5}$$

this expression was called the "ordered derivative" by Werbos (1974). The difference between the ordered derivative and the ordinary partial derivative lies in the way we view the function to be differentiated. For an internal node output $x_{l,i}$, the partial derivative $\partial^+ E_p / \partial x_{l,i}$ is equal to zero, since E_p does not depend on $x_{l,i}$ directly. However, it is obvious that E_p does depend on $x_{l,i}$ indirectly, since a change in $x_{l,i}$ will propagate through indirect paths to the output layer and thus produce a corresponding change in the value of E_p .

The error signal for the ith output node (at layer L) can be calculated directly:

$$\varepsilon_{L,i} = \frac{\partial^+ E_p}{\partial x_{L,i}} = \frac{\partial E_p}{\partial x_{L,i}} \tag{3.6}$$

This is equal to $\varepsilon_{L,i} = -2(d_i - x_{L,i})$ if E_p is defined as in Equation (3.4). For the internal node at the ith position of layer l, the error signal can be derived by the chain rule of differential calculus:

$$\varepsilon_{l,i} = \underbrace{\frac{\partial^+ E_p}{\partial x_{l,i}}}_{\substack{\text{error signal} \\ \text{at layer l}}} = \sum_{m=1}^{N(l+1)} \underbrace{\frac{\partial^+ E_p}{\partial x_{l+1,m}}}_{\substack{\text{error signal} \\ \text{at layer l+1}}} \frac{\partial f_{l+1,m}}{\partial x_{l,i}} = \sum_{m=1}^{N(l+1)} \varepsilon_{l+1,m} \frac{\partial f_{l+1,m}}{\partial x_{l,i}} \tag{3.7}$$

where $0 \le l \le L-1$. That is, the error signal of an internal node at layer l can be expressed as a linear combination of the error signal of the nodes at layer l+1. Therefore, for any l and i, we can find $\varepsilon_{l,i}$ by first applying Equation (3.6) once to get error signals at the output layer, and then applying Equation (3.7) iteratively until we reach the desired layer l. The underlying procedure is called backpropagation since the error signals are obtained sequentially from the output layer back to the input layer.

The gradient vector is defined as the derivative of the error measure with respect to each parameter, so we have to apply the chain rule again to find the gradient vector. If α is a parameter of the ith node at layer l, we have

$$\frac{\partial^+ E_p}{\partial \alpha} = \frac{\partial^+ E_p}{\partial x_{l,i}} \frac{\partial f_{l,i}}{\partial \alpha} = \varepsilon_{l,i} \frac{\partial f_{l,i}}{\partial \alpha} \qquad (3.8)$$

The derivative of the overall error measure E with respect to α is

$$\frac{\partial^+ E}{\partial \alpha} = \sum_{p=1}^{P} \frac{\partial^+ E_p}{\partial \alpha} \qquad (3.9)$$

Accordingly, for simple steepest descent (for minimization), the update formula for the generic parameter α is

$$\Delta\alpha = - \eta \frac{\partial^+ E}{\partial \alpha} \qquad (3.10)$$

in which η is the "learning rate", which can be further expressed as

$$\eta = \frac{k}{\sqrt{\sum_\alpha (\partial E / \partial \alpha)^2}} \qquad (3.11)$$

where k is the "step size", the length of each transition along the gradient direction in the parameter space.

There are two types of learning paradigms that are available to suit the needs for various applications. In "off-line learning" (or "batch learning"), the update formula for parameter α is based on Equation (3.9) and the update action takes place only after the whole training data set has been presented-that is, only after each "epoch" or "sweep". On the other hand, in "on-line learning" (or "pattern-by-pattern learning"), the parameters are updated immediately after each input-output pair has been presented, and the update formula is based on Equation (3.8). In practice, it is possible to combine these two learning modes and update the parameter after k training data entries have been presented, where k is between 1 and P and it is sometimes referred to as the "epoch size".

3.1.2 Backpropagation multilayer perceptrons

Artificial neural networks, or simply "neural networks" (NNs), have been studied for more than three decades since Rosenblatt first applied single-layer "perceptrons" to pattern classification learning (Rosenblatt, 1962). However, because Minsky and Papert pointed out that single-layer systems were limited and expressed pessimism over multilayer systems, interest in NNs dwindled in the 1970s (Minsky & Papert, 1969). The recent resurgence of interest in the field of NNs has been inspired by new developments in NN learning algorithms (Fahlman & Lebiere, 1990), analog VLSI circuits, and parallel processing techniques (Lippmann, 1987).

Quite a few NN models have been proposed and investigated in recent years. These NN models can be classified according to various criteria, such as their learning methods (supervised versus unsupervised), architectures (feedforward versus recurrent), output types (binary versus continuous), and so on. In this section, we confine our scope to modelling problems with desired input-output data sets, so the resulting networks must have adjustable parameters that are updated by a supervised learning rule. Such networks are often referred to as "supervised learning" or "mapping networks", since we are interested in shaping the input-output mappings of the network according to a given training data set.

A backpropagation "multilayer perceptron" (MLP) is an adaptive network whose nodes (or neurons) perform the same function on incoming signals; this node function is usually a composite of the weighted sum and a differentiable non-linear activation function, also known as the "transfer function". Figure 3.3 depicts three of the most commonly used activation functions in backpropagation MLPs:

Logistic function: $\qquad f(x) = \dfrac{1}{1 + e^{-x}}$

Hyperbolic tangent function: $f(x) = \tan h\,(x/2) = \dfrac{1 - e^{-x}}{1 + e^{-x}}$

Identity function: $\qquad f(x) = x$

Both the hyperbolic tangent and logistic functions approximate the signum and step function, respectively, and provide smooth, nonzero derivatives with respect to input signals. Sometimes these two activation functions are referred to as "squashing functions" since the inputs to these functions are squashed to the range [0,1] or [-1,1]. They are also called "sigmoidal functions" because their s-shaped curves exhibit smoothness and asymptotic properties.

Backpropagation MLPs are by far the most commonly used NN structures for applications in a wide range of areas, such as pattern recognition, signal processing, data compression and automatic control. Some of the well-known instances of applications include NETtalk (Sejnowski & Rosenberg, 1987), which trained an MLP to pronounce English text, Carnegie Mellon University's ALVINN (Pomerleau, 1991), which used an MLP for steering an autonomous vehicle; and optical character recognition (Sakinger, Boser, Bromley, Lecun & Jackel, 1992). In the following lines, we derive the backpropagation learning rule for MLPs using the logistic function.

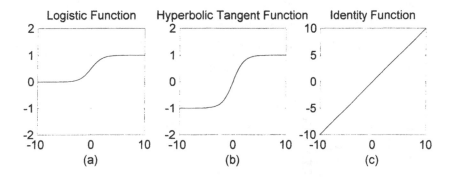

Figure 3.3 Activation functions for backpropagation MLPs: (a) logistic function; (b) hyperbolic function; (c) identity function.

The "net input" $\bar{x}$ of a node is defined as the weighted sum of the incoming signals plus a bias term. For instance, the net input and output of node j in Figure 3.4 are

$$\bar{x}_j = \sum_i w_{ij} x_i + w_j \ , \qquad\qquad (3.12)$$

$$x_j = f(\bar{x}_j) = \frac{1}{1 + \exp(-\bar{x}_j)} \ ,$$

where x_i is the output of node i located in any one of the previous layers, w_{ij} is the weight associated with the link connecting nodes i and j, and w_j is the bias of node j. Since the weights w_{ij} are actually internal parameters associated with each node j, changing the weights of a node will alter the behavior of the node and in turn alter the behavior of the whole backpropagation MLP.

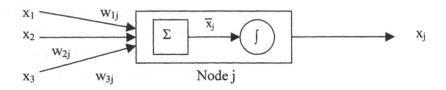

Figure 3.4 Node j of a backpropagation MLP.

Figure 3.5 shows a three-layer backpropagation MLP with three inputs to the input layer, three neurons in the hidden layer, and two output neurons in the output layer. For simplicity, this MLP will be referred to as a 3-3-2 network, corresponding to the number of nodes in each layer.

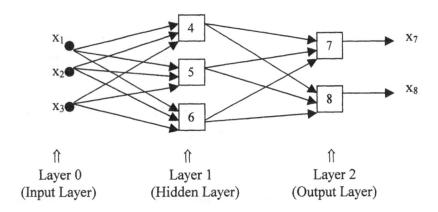

Figure 3.5 A 3-3-2 backpropagation MLP.

The "backward error propagation", also known as the "backpropagation" (BP) or the "generalized data rule" (GDR), is explained next. First, a squared error measure for the pth input-output pair is defined as

$$E_p = \sum_k (d_k - x_k)^2 \tag{3.13}$$

where d_k is the desired output for node k, and x_k is the actual output for node k when the input part of the pth data pair presented. To find the gradient vector, an error term ε_i is defined as

$$\bar{\varepsilon}_i = \frac{\partial^+ E_p}{\partial x_i} \tag{3.14}$$

By the chain rule, the recursive formula for ε_i can be written as

$$\bar{\varepsilon}_i = \begin{cases} -2(d_i - x_i) \dfrac{\partial x_i}{\partial \bar{x}_i} = -2(d_i - x_i) x_i (1 - x_i) & \text{if node i is a output node} \\[2em] \dfrac{\partial x_i}{\partial \bar{x}_i} \sum_{j, i<j} \dfrac{\partial^+ E_p}{\partial \bar{x}_j} \dfrac{\partial \bar{x}_j}{\partial x_i} = x_i (1 - x_i) \sum_{j, i<j} \bar{\varepsilon}_j w_{ij} & \text{otherwise} \end{cases} \tag{3.15}$$

where w_{ij} is the connection weight from node i to j; and w_{ij} is zero if there is no direct connection. Then the weight update w_{ki} for on-line (pattern-by-pattern) learning is

$$\Delta w_{ki} = -\eta \frac{\partial^+ E_p}{\partial w_{ki}} = -\eta \frac{\partial^+ E_p}{\partial \bar{x}_i} \frac{\partial \bar{x}_i}{\partial w_{ki}} = -\eta \bar{\varepsilon}_i x_k \tag{3.16}$$

where η is a learning rate that affects the convergence speed and stability of the weights during learning.

For off-line (batch) learning, the connection weight w_{ki} is updated only after presentation of the entire data set, or only after an "epoch":

$$\Delta w_{ki} = -\eta \frac{\partial^+ E}{\partial w_{ki}} = -\eta \sum_p \frac{\partial^+ E_p}{\partial w_{ki}} \tag{3.17}$$

or, in vector form,

$$\Delta \mathbf{w} = -\eta \frac{\partial^+ E}{\partial \mathbf{w}} = -\eta \nabla_{\mathbf{w}} E \tag{3.18}$$

where $E = \sum_p E_p$. This corresponds to a way of using the true gradient direction based on the entire data set.

The approximation power of backpropagation MLPs has been explored by some researchers. Yet there is very little theoretical guidance for determining network size in terms of, say, the number of hidden nodes and hidden layers it should contain. Cybenko (1989) showed that a backpropagation MLP, with one hidden layer and any fixed continuous sigmoidal non-linear function, can approximate any continuous function arbitrarily well on a compact set. When used as a binary-valued neural network with the step activation function, a backpropagation MLP with two hidden layers can form arbitrary complex decision regions to separate different classes, as Lippmann (1987) pointed out. For function approximation as well as data classification, two hidden layers may be required to learn a piecewise-continuous function (Masters, 1993).

3.2 Adaptive Neuro-Fuzzy Inference Systems

In this section, we describe a class of adaptive networks that are functionally equivalent to fuzzy inference systems (Kosko, 1992). The architecture is referred to as ANFIS, which stands for "adaptive network-based fuzzy inference system". We describe how to decompose the parameter set to facilitate the hybrid learning rule for ANFIS architectures representing both the Sugeno and Tsukamoto fuzzy models.

3.2.1 ANFIS architecture

A fuzzy inference system consists of three conceptual components: a fuzzy rule base, which contains a set of fuzzy if-then rules; a database, which defines the membership functions used in the fuzzy rules; and a reasoning mechanism, which performs the inference procedure upon the rules to derive a reasonable output or conclusion (Kandel, 1992). For simplicity, we assume that the fuzzy inference system under consideration has two inputs x and y and one output z. For a first-order Sugeno fuzzy model (Sugeno & Kang, 1988), a common rule set with two fuzzy if-then rules is the following:

Rule 1: If x is A_1 and y is B_1, then $f_1 = p_1x + q_1y + r_1$,

Rule 2: If x is A_2 and y is B_2, then $f_2 = p_2x + q_2y + r_2$,

Figure 3.6 (a) illustrates the reasoning mechanism for this Sugeno model; the corresponding equivalent ANFIS architecture is as shown in Figure 3.6 (b), where nodes of the same layer have similar functions, as described next. (Here we denote the output of the ith node in layer l as $0_{l,i}$).

Layer 1: Every node i in this layer is an adaptive node with a node function

$$0_{l,i} = \mu_{Ai}(x), \text{ for } i = 1, 2 ,$$
$$0_{l,i} = \mu_{Bi-2}(y), \text{ for } i = 3, 4 , \tag{3.19}$$

where x (or y) is the input to node i and A_i (or B_{i-2}) is a linguistic label (such as "small" or "large") associated with this node. In other words, $0_{l,i}$ is the membership grade of a fuzzy set A and it specifies the degree to which the given input x (or y) satisfies the quantifier A. Here the membership function for A can be any appropriate parameterized membership function, such as the generalized bell function:

$$\mu_A(x) = \frac{1}{1 + | (x-c_i)/a_i |^{2bi}} \tag{3.20}$$

where $\{a_i , b_i , c_i\}$ is the parameter set. As the values of these parameters change, the bell-shaped function varies accordingly, thus exhibiting various forms of membership functions for a fuzzy set A. Parameters in this layer are referred to as "premise parameters".

Layer 2: Every node in this layer is a fixed node labeled Π, whose output is the product of all incoming signals:

$$0_{2,i} = w_i = \mu_{Ai}(x) \mu_{Bi}(y), \quad i = 1, 2 . \tag{3.21}$$

Each node output represents the firing strength of a fuzzy rule.

Layer 3: Every node in this layer is a fixed node labeled N. The ith node calculates the ratio of the ith rule's firing strength to the sum of all rules' firing strengths:

$$0_{3,i} = \overline{w}_i = w_i / (w_1 + w_2) , \quad i = 1,2. \tag{3.22}$$

For convenience, outputs of this layer are called "normalized firing strengths".

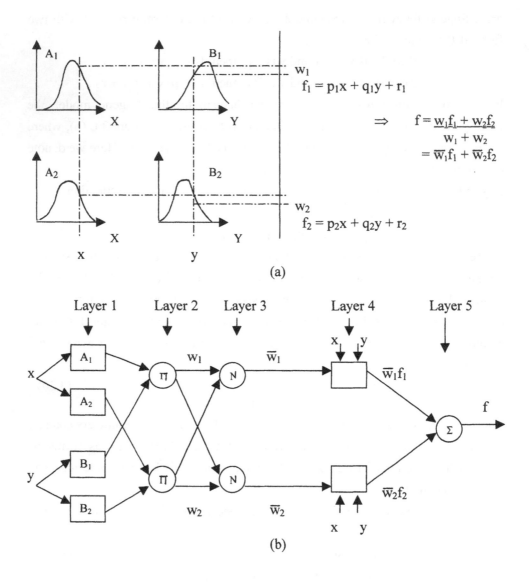

Figure 3.6 (a) A two-input Sugeno fuzzy model with 2 rules; (b) equivalent
        ANFIS architecture (adaptive nodes shown with a square and fixed
        nodes with a circle).

Layer 4: Every node i in this layer is an adaptive node with a node function
$$O_{4,i} = \overline{w}_i\, f_i = \overline{w}_i\, (\, p_i x + q_i y + r_i\,) \quad , \tag{3.23}$$
where $\overline{w}_i$ is a normalized firing strength from layer 3 and $\{p_i\, ,\, q_i\, ,\, r_i\,\}$ is the parameter set of this node. Parameters in this layer are referred to as "consequent parameters".

Layer 5: The single node in this layer is a fixed node labeled Σ, which computes the overall output as the summation of all incoming signals:
$$\text{overall output} = O_{5,i} = \sum_i \overline{w}_i\, f_i = \frac{\sum_i w_i\, f_i}{\sum_i w_i} \tag{3.24}$$

Thus we have constructed an adaptive network that is functionally equivalent to a Sugeno fuzzy model. We can note that the structure of this adaptive network is not unique; we can combine layers 3 and 4 to obtain an equivalent network with only four layers. In the extreme case, we can even shrink the whole network into a single adaptive node with the same parameter set. Obviously, the assignment of node functions and the network configuration are arbitrary, as long as each node and each layer perform meaningful and modular functionalities.

The extension from Sugeno ANFIS to Tsukamoto ANFIS is straightforward, as shown in Figure 3.7, where the output of each rule (f_i, i = 1, 2) is induced jointly by a consequent membership function and a firing strength.

3.2.2 Learning algorithm

From the ANFIS architecture shown in Figure 3.6 (b), we observe that when the values of the premise parameters are fixed, the overall output can be expressed as a linear combination of the consequent parameters. Mathematically, the output f in Figure 3.6 (b) can be written as
$$f = \frac{w_1}{w_1 + w_2}\, f_1 + \frac{w_2}{w_1 + w_2}\, f_2 \tag{3.25}$$
$$= \overline{w}_1\, (\, p_1 x + q_1 y + r_1\,) + \overline{w}_2\, (p_2 x + q_2 y + r_2)$$
$$= (\overline{w}_1 x)\, p_1 + (\overline{w}_1 y)\, q_1 + (\overline{w}_1)\, r_1 + (\overline{w}_2 x)\, p_2 + (\overline{w}_2 y)\, q_2 + (\overline{w}_2)\, r_2$$

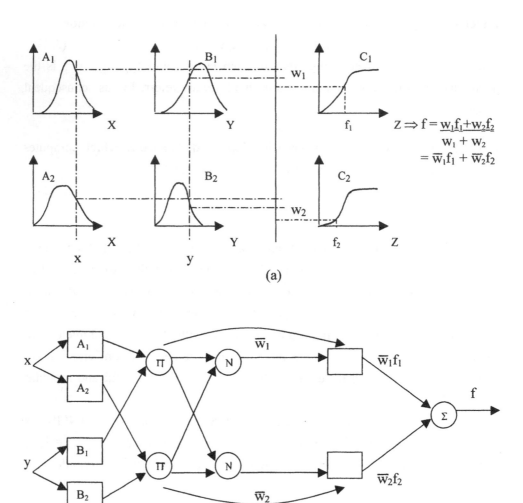

Figure 3.7 (a) A two-input Tsukamoto fuzzy model with two rules;
(b) equivalent ANFIS architecture.

which is linear in the consequent parameters p_1, q_1, r_1, p_2, q_2, and r_2. From this observation, we can use a hybrid learning algorithm for parameter estimation in this kind of models (Jang, 1993). More specifically, in the forward pass of the hybrid learning algorithm, node outputs go forward until layer 4 and the consequent parameters are identified by the least-squares method. In the backward pass, the error signals propagate backward and the premise parameters are updated by gradient descent.

It has been shown (Jang, 1993) that the consequent parameters identified in this manner are optimal under the condition that the premise parameters are fixed. Accordingly, the hybrid approach converges much faster since it reduces the search space dimensions of the original pure backpropagation method. For Tsukamoto ANFIS, this can be achieved if the membership function on the consequent part of each rule is replaced by a piecewise linear approximation with two consequent parameters.

3.3 Neuro-Fuzzy Control

The original purpose of fuzzy logic control, as proposed in Mamdani's paper in 1975, was to mimic the behavior of a human operator able to control a complex plant satisfactorily. The complex plant in question could be a chemical reaction process, a subway train, or a traffic signal control system. After more than 20 years, the ultimate goal of fuzzy controllers remains the same-that is, to automate an entire control process by replacing a human operator with a fuzzy controller made up of computer software/hardware.

To construct a fuzzy controller, we need to perform "knowledge acquisition", which takes a human operator's knowledge about how to control a system and generates a set of fuzzy if-then rules as the backbone for a fuzzy controller that behaves like the original human operator. Usually we can obtain two types of information from a human operator: "linguistic information" and "numerical information".

Linguistic information: An experienced human operator can usually summarize his or her reasoning process in arriving at final control actions or decisions as a set of fuzzy if-then rules with imprecise but roughly correct membership functions;

this corresponds to the linguistic information supplied by human experts, which is obtained via a lengthy interview process plus a certain amount of trial and error.

Numerical information: When a human operator is working, it is possible to record the sensor data observed by the human and the human's corresponding actions as a set of desired input-output data pairs. This data set can be used as training data in constructing a fuzzy controller.

Prior to the emergence of neuro-fuzzy approaches, most design methods used only linguistic information to build fuzzy controllers; this approach is not easily formalized and is more of an art than an engineering practice. Following this approach usually involves manual trial-and-error processes to fine-tune the membership functions. Successful fuzzy control applications based on linguistic information plus trial-and-error tuning include steam engine and boiler control (Mamdani & Assilian, 1975), Sendai subway systems (Yasunobu & Miyamoto, 1985), nuclear reaction control (Bernard, 1988), automobile transmission control (Kasai & Morimoto, 1988), aircraft control (Chiu, Chand, Moore & Chaudhary, 1991), and many others.

Now, with learning algorithms, we can take further advantage of the numerical information (input-output data pairs) and refine the membership functions in a systematic way. In other words, we can use linguistic information to identify the structure of a fuzzy controller, and then use numerical information to identify the parameters such that the fuzzy controller can reproduce the desired action more accurately.

3.3.1 Inverse learning

The development of "inverse learning" (Widrow & Stearns, 1985) for designing neuro-fuzzy controllers involves two phases. In the learning phase, an on-line or off-line technique is used to model the inverse dynamics of the plant. The obtained neuro-fuzzy model, which represents the inverse dynamics of the plant, is then used to generate control actions in the application phase. These two phases, can proceed simultaneously, hence this design method fits in perfectly with the classical adaptive control scheme.

By assuming that the order of the plant (that is, the number of state variables) is known and all state variables are measurable, we have

$$\mathbf{x}(k+1) = \mathbf{f}(\mathbf{x}(k), u(k)) \tag{3.26}$$

where $\mathbf{x}(k+1)$ is the state at time $k+1$, $\mathbf{x}(k)$ is the state at time k, and u(k) is the control signal at time k (assuming for simplicity that u(k) is a scalar). Similarly, the state at time $k+2$ is expressed as

$$\mathbf{x}(k+2) = \mathbf{f}(\mathbf{x}(k+1), u(k+1)) = \mathbf{f}(\mathbf{f}(\mathbf{x}(k), u(k)), u(k+1)) \tag{3.27}$$

In general, we have

$$\mathbf{x}(k+n) = \mathbf{F}(\mathbf{x}(k), \mathbf{U}) \tag{3.28}$$

where n is the order of the plant, $\mathbf{F}$ is a multiple composite function of $\mathbf{f}$, and $\mathbf{U}$ is the control actions from k to $k+n-1$, which is equal to

$$[u(k), u(k+1), ..., u(k+n-1)]^T$$

The preceding equation points out the fact that given the control input u from time k to $k+n-1$, the state of the plant will move from $\mathbf{x}(k)$ to $\mathbf{x}(k+n)$ in exactly n time steps. Furthermore, we assume that the inverse dynamics of the plant do exist, that is, $\mathbf{U}$ can be expressed as an explicit function of $\mathbf{x}(k)$ and $\mathbf{x}(k+n)$:

$$\mathbf{U} = \mathbf{G}(\mathbf{x}(k), \mathbf{x}(k+n)) \tag{3.29}$$

This equation essentially says that there exists a unique input sequence $\mathbf{U}$, specified by mapping $\mathbf{G}$, that can drive the plant from state $\mathbf{x}(k)$ to $\mathbf{x}(k+n)$ in n time steps. The problem now becomes how to find the inverse mapping $\mathbf{G}$.

Although the inverse mapping $\mathbf{G}$ in Equation (3.29) exists by assumption, it does not always have an analytically closed form. Therefore, instead of looking for methods of solving Equation (3.29) explicitly, we can use an adaptive network or ANFIS with 2n inputs and n outputs to approximate the inverse mapping $\mathbf{G}$ according to the generic training data pairs

$$[\mathbf{x}(k)^T, \mathbf{x}(k+n)^T; \mathbf{U}^T] \tag{3.30}$$

Figure 3.8 illustrates the situation in which n is equal to 1. Figure 3.8 (a) shows a plant block in which the plant output x(k+1) is a function of a previous state x(k) and input u(k); we use z^{-1} block to represent the unit-time delay operator. Figure 3.8 (b) is the block diagram during the training phase; Figure 3.8 (c) is the block diagram during the application phase.

Assume that the adaptive network truly imitates the input-output mapping of the inverse dynamics $\mathbf{G}$. Then, given the current state $\mathbf{x}(k)$ and the desired future state $\mathbf{x}_d(k+n)$, the adaptive network will generate an estimated $\hat{\mathbf{U}}$:

$$\hat{\mathbf{U}} = \hat{\mathbf{G}}(\mathbf{x}(k), \mathbf{x}_d(k+n)) \tag{3.31}$$

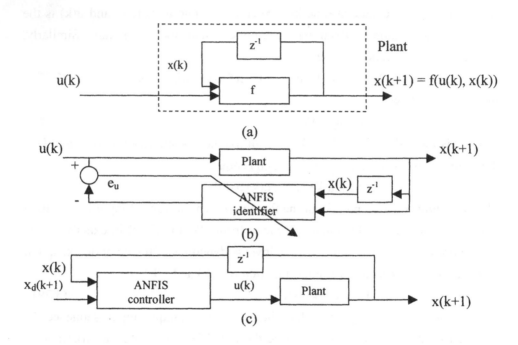

Figure 3.8 Block diagram for the inverse learning method: (a) plant block; (b) training phase; (c) application phase

After n steps, this control sequence can bring the state $\mathbf{x}(k)$ to the desired state $\mathbf{x}_d(k+n)$, assuming that the adaptive network function $\hat{\mathbf{G}}$ is exactly the same as the inverse mapping $\mathbf{G}$. This application phase is shown in the block diagram of Figure 3.8 (b). If the future desired state $\mathbf{x}_d(k+n)$ is not available in advance, we can use the current desired state $\mathbf{x}_d(k)$ in Figure 3.8 (b). This implies that the current desired state will appear after n time steps and the whole system behaves like a pure n-step time delay system.

When $\hat{G}$ is not close to G, the control sequence U cannot bring the state to $x_d(k+n)$ in exactly the next n time step. As more data pairs are used to refine the parameters in the adaptive network, $\hat{G}$ will become closer to G and the control will be more and more accurate as the training process goes on.

For off-line applications, we have to collect a set of training data pairs and then train the adaptive network in the batch mode. For on-line applications to deal with time-varying systems, the control actions in Equation (3.31) are generated every n time steps while on-line learning occurs at every time step. Alternatively, we can generate the control sequence at every time step and apply only the first component to the plant. Figure 3.9 is a block diagram for on-line learning when n is equal to 1. The dashed line in the figure indicates that the two ANFIS blocks are exact duplicates of each other. (For simplicity, we have removed the unit-time delay operator from this figure).

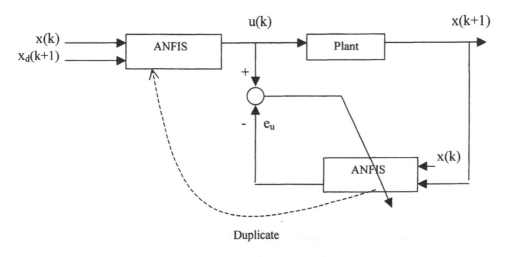

Figure 3.9 Block diagram for on-line inverse learning

3.3.2 Specialized learning

A major problem with inverse learning is that an inverse model does not always exist for a given plant. Moreover, inverse learning is an indirect approach that tries to minimize the network output error instead of the overall system error (defined as the difference between desired and actual trajectories). "Specialized learning" (Psaltis, Sideris & Yamamura, 1988) is an alternative method that tries to minimize the system error directly by backpropagating error signals through the plant block. The price that we pay is that we need to know more about the plant under consideration.

Figure 3.10 illustrates the most basic type of specialized learning, Figure 3.10 (a) is the plant block (assuming its order is 1), and Figure 3.10 (b) indicates the training of the ANFIS controller. The ANFIS parameters are updated to reduce the system error $e_x(k)$, which is defined as the difference between the system's output $x(k)$ and the desired output $x_d(k)$.

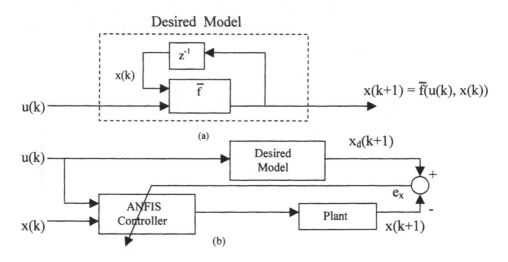

Figure 3.10 (a) Desired model block; (b) specialized learning
with model referencing

To be more specific, let the plant dynamics be specified by
$$x(k+1) = f(x(k), v(k))$$
and the ANFIS output be denoted as
$$\hat{v}(k) = F(x(k), u(k), \theta) \tag{3.32}$$

where θ is a parameter vector to be updated. If we set the ANFIS output as the plant's input, then $v(k) = \hat{v}(k)$ and we have a closed-loop system specified by

$$\mathbf{x}(k+1) = \mathbf{f}(\mathbf{x}(k), F(\mathbf{x}(k), u(k), \theta))$$

The objective of specialized learning is to minimize the difference between the closed-loop system and the desired model. Hence we can define an error measure:

$$J(\theta) = \sum_k \|\mathbf{f}(\mathbf{x}(k), F(\mathbf{x}(k), u(k), \theta)) - \mathbf{x}_d(k+1)\|^2 \qquad (3.33)$$

We can use backpropagation or steepest descent to update θ to minimize the above error measure. To find the derivative of $J(\theta)$ with respect to θ, we need to know the derivative of $\mathbf{f}$ with respect to its second argument. In other words, to backpropagate error signals through the plant block in Figure 3.10 (b), we need to know the "Jacobian matrix" of the plant, where the element at row i and column j is equal to the derivative of the plant's ith output with respect to its jth input. This usually implies that we need a model for the plant and the Jacobian matrix obtained from the model, which could be a neural network, an ANFIS, or another appropriate mathematical description of the plant.

For a single-input plant, if the Jacobian matrix is not easily found directly, a crude estimate can be obtained by approximating it directly from the changes in the plant's input and output(s) during two consecutive time instants. Other methods that aim at using an approximate Jacobian matrix to achieve the same learning effects can be found in Chen and Pao (1989).

It is not always convenient to specify the desired plant output $\mathbf{x}_d(k)$ at every time instant k. As a standard approach in model reference adaptive Control, the desired behavior of the overall system can be implicitly specified by a model that is able to achieve the control goal satisfactorily. Let the desired model be specified by

$$\mathbf{x}_d(k+1) = \overline{\mathbf{f}}(\mathbf{x}(k), u(k))$$

Then the error measure in Equation (3.33) becomes

$$J(\theta) = \sum_k \|\mathbf{f}(\mathbf{x}(k), F(\mathbf{x}(k), u(k), \theta)) - \overline{\mathbf{f}}(\mathbf{x}(k), u(k))\|^2 \qquad (3.34)$$

Again, we still need the Jacobian matrix of the plant to do backpropagation.

Note that the ANFIS controller in Equation (3.32) represents the most general situation. More commonly, the ANFIS controller is a function of $\mathbf{x}(k)$ and θ only and the input to the plant v(k) is expressed as the difference between the command signal u(k) and ANFIS output, as follows:

$$\hat{v}(k) = u(k) - F(\mathbf{x}(k), \theta) \,.$$

3.4 Adaptive Model-Based Neuro-Control

This section briefly reviews various approaches in current adaptive neuro-control design (Odmivar & Elliot, 1997). Although there are other ways to classify these approaches (e.g., Hunt, Sbarbaro, Zbikowski & Gawthrop, 1992) this section nevertheless adopts one similar to adaptive control theory: 1) indirect neuro-control and 2) direct neuro-control.

In the indirect neuro-control scheme, a neural network does not send a control signal "directly" to the process. Instead, a neural network is often used as an indirect process characteristics indicator. This indicator can be a process model that mimics the process behavior or a controller auto-tuner that produces appropriate controller settings based upon the process behavior. In this category, the neuro-control approaches can be roughly distinguished as follows: 1) neural network model-based control, 2) neural network inverse model-based control, and 3) neural network auto-tuner development.

In the direct neuro-control scheme, a neural network is employed as a feedback controller, and it sends control signals "directly" to the process. Depending on the design concept, the direct neuro-control approaches can be categorized into: 1) controller modelling, 2) model-free neuro-control design, 3) model-based neuro-control design, and 4) robust model-based neuro-control design.

Regardless of these distinctions, a unifying framework for neuro-control is to view neural network training as a non-linear optimization problem,

$$\text{NN: } \min_{w} J(w) \tag{3.35}$$

in which one tries to find an optimal representation of the neural network that minimizes an objective function J over the network weight space w. Here, NN indicates that the optimization problem formulation involves a neural network.

The role a neural network plays in the objective function is then a key to distinguishing the various neuro-control design approaches.

3.4.1 Indirect Neuro-Control

The most popular control system application of neural networks is to use a neural network as an input-output process model. This approach is a data-driven supervised learning approach, i.e., the neural network attempts to mimic an existing process from being exposed to the process data (see Figure 3.11). The most commonly adopted model structure for such a purpose is the non-linear auto-regressive and moving average with exogenous inputs (known as NARMAX) model or a simpler NARX (Su, McAvoy & Werbos, 1992). Alternatively, one can choose to identify a continuous-time model with a dynamic neural network. Regardless of the model structure and the control strategy, the neuro-control design in this case can be conceptually stated as follows:

$$\text{NN: } \min_{w} F\{\, y_p - y_n(w, ...) \,\} \tag{3.36}$$

where y_p stands for plant/process output, y_n for neural network output, and w for neural network weights. Here F is a functional that measures the performance of the optimization process. It is usually an integral or sum of the prediction errors between y_p and y_n. For example, in this model development stage, process inputs and outputs $\{u_p, y_p\}$ are collected over a finite period of time and used for neural network training.

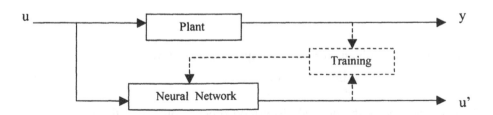

Figure 3.11 Neural Network as a black-box model of a process

At the implementation stage, nevertheless, the neural network model cannot be used alone. It must be incorporated with a model-based control scheme. In the chemical process industry, for example, a neural network is usually employed in a non-linear model predictive control (MPC) scheme (Su & McAvoy, 1993). Figure 3.12 illustrates the block diagram of an MPC control system. In fact, the MPC control is also an optimization problem.

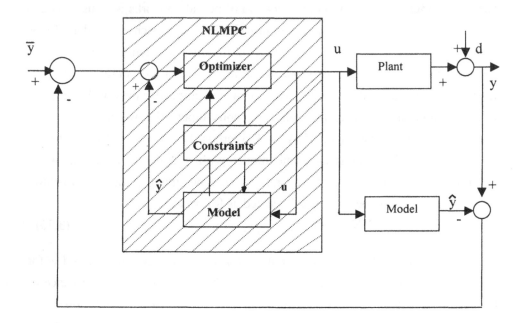

Figure 3.12 Neural network model with non-linear model
predictive control (MPC)

The optimization problem here can be expressed as follows:

$$\min_{u} F' \{ y^* - y_n (u, ...) \} \tag{3.37}$$

where y^* designates the desired close-loop process output, u the process/model input or control signal, and y_n the predicted process output (by the neural network model). Here F' stands for an objective function that evaluates the closed-loop performance. For example, the optimization problem in the implementation stage is usually as follows:

$$\min_{u} \; \Sigma_{t} \; \| \, y^*(t) - y_n(t) - d(t) \, \|^2 \;\; , \;\; y_n = \aleph \, (u, \, ...) \qquad (3.38)$$

where $y^*(t)$ stands for desired set point trajectory and $d(t)$ for estimated disturbance. This optimization is performed repeatedly at each time interval during the course of feedback control. Although the constrains are not particularly of interest in the discussion, one advantage of this indirect control design approach over the direct ones is that the constraints can be incorporated when solving the above optimization problem.

In some cases, a certain degree of knowledge, about the process might be available, such as model structure or particular physical phenomena that are well understood. In this case, a full black-box model might not be most desirable. For example, if the structure of the process model is available, values for the associated parameters can be determined by a neural network. Examples of these parameters can be time constants, gains, and delays or physical parameters such as diffusion rates and heat transfer coefficients. When model structure is not known a priori, neural networks can be trained to select elements of a model structure from a predetermined set. Lastly, in other cases where model structure is partially known, neural networks can also be integrated with such a partial model so that the process can be better model (see Figure 3.13).

For illustration purposes, the parametric or partial neural network modelling problem can be formulated as follows:

$$\text{NN:} \; \min_{w} \; F \, \{ \, y_p - y_m \, (\theta, \, ... \,) \, \} \;\; , \;\; \theta = \aleph \, (w, \, ...) \qquad (3.39)$$

where y_m is the predicted output from the model and θ stands for the process parameters, model structural information and other elements required to complete the model. Notice the only difference between Equation (3.39) and Equation (3.36) is that y_m replaces y_n. From a model-based control standpoint, this approach is essentially identical to the full black-box neural network model except that the neural network does not directly mimic the process behavior.

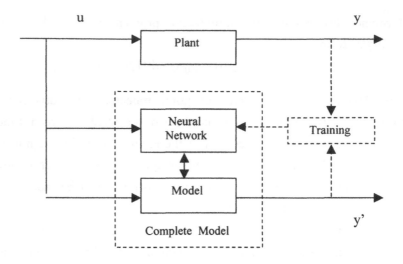

Figure 3.13 A neural network can be a parameter estimator, model structure selector, or a partial element of a physical model

A neural network can be trained to develop an inverse model of the plant. The network input is the process output, and the network output is the corresponding process input (see Figure 3.14). In general, the optimization problem can be formulated as

$$\text{NN: } \min_{w} F \{ u^*_{p-1} - u_n (w, \dots) \} \qquad (3.40)$$

where u^*_{p-1} is the process inputs. Typically, the inverse model is a steady state/static model, which can be used for feedforward control. Given a desired process set point y^*, the appropriate steady-state control signal u^* for this set point can be immediately known:

$$u^* = \aleph (y^*, \dots) \qquad (3.41)$$

Successful applications of inverse modelling are discussed in (Miller, Sutton & Werbos, 1995). Obviously, an inverse model exists only when the process behaves monotonically as a "forward" function at steady state.

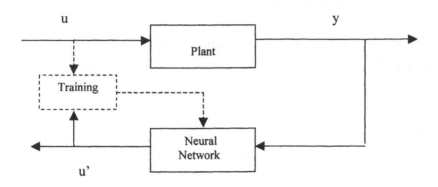

Figure 3.14 A neural network inverse model

As in the previous case where neural networks can be used to estimate parameters of a known model, they can also be used to estimate tuning parameters of a controller whose structure is known a priori. A controller's tuning parameter estimator is often referred to as an autotuner. The optimization problem in this case can be formulated as follows:

$$\text{NN: } \min_{w} F \{ \eta^* - \eta_n (w, \dots) \} \qquad (3.42)$$

where η^* denotes the controller parameters as targets and η_n stands for the predicted values by the neural network. Network input can comprise sampled process data or features extracted from it. However, these parameters η cannot be uniquely determined from the process characteristics. They also depend on the desired closed-loop control system characteristics. Usually, the controller parameters are solutions to the following closed-loop control optimization:

$$\min_{\eta} F' \{ y^* - y_{p/m} (u, \dots) \} \quad ; \quad u = C(\eta, \dots) \qquad (3.43)$$

where C is a controller with a known structure. Here, $y_{p/m}$ denotes that either a process or a model can be employed in this closed-loop control in order to find the target controller C.

3.4.2 Direct Neuro-Control

Among the four direct neuro-control schemes, the simplest for neuro-controller development is to use a neural network to model an existing controller (see Figure 3.15). The input to the existing controller is the training input to the network and the controller output serves as the target. This neuro-control design can be formulated as follows:

$$\text{NN: } \min_{w} F \{u^*_c - u_n (w, \dots)\} \tag{3.44}$$

where u^*_c is the output of an existing controller C^*. Usually, the existing controller C^* can be a human operator or it can be obtained via

$$\min_{c} F' \{y^* - y_{p/m} (u, \dots)\} \quad ; \quad u = C(\dots) \tag{3.45}$$

Like a process model, a controller is generally a dynamical system and often comprises integrators or differentiators. If a feedforward network is used to model the existing controller, dynamical information must be explicitly provided as input to the network.

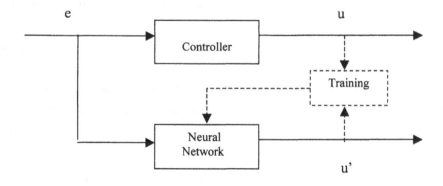

Figure 3.15 The simplest approach to neuro-control is to use a
neural network to model an existing controller

While the benefits of this approach may be apparent when the existing controller is a human, its utility may be limited. It is applicable only when an existing controller is available, which is the case in many applications. Staib & Staib (1992) discuss how it can be effective in a multistage training process.

In the absence of an existing controller, some researchers have been inspired by the way a human operator learns to "control/operate" a process with little or no detailed knowledge of the process dynamics. Thus they have attempted to design controllers that by adaptation and learning can solve difficult control problems in the absence of process models and human design effort. In general, this model-free neuro-control can be stated as:

$$\text{NN: } \min_{w} F \{y^* - y_p (u, ...)\} \quad , \quad u = \aleph (w, ...) \qquad (3.46)$$

where y_p is the output from the plant. The key feature of this direct adaptation control approach is that a process model is neither known in advance nor explicitly developed during control design. Figure 3.16 is a typical representation of this class of control design.

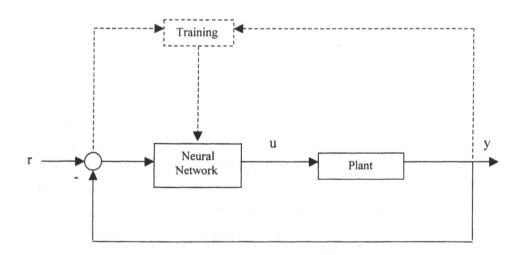

Figure 3.16 The model-free control design concept

The first work in this area was the "adaptive critic" algorithm proposed by Barto et al. (1983). Such an algorithm can be seen as an approximate version of dynamic programming. In this work, they posed a well-known cart-pole balancing problem and demonstrated their design concept. In this class of control design, limited/poor information is often adopted as an indication of performance criteria. For example, the objective is the cart-pole balancing problem is simply to maintain the pole in a near-upright balanced position for as long as possible. The instructional feedback is limited to a "failure" signal when the controller fails to hold the pole in an upright position. The cart-pole problem has become a popular test-bed for explorations of the model-free control design concept.

Despite its historical importance and intuitive appeal, model-free adaptive neuro-control is not appropriate for most real-world applications. The plant is most likely out of control during the learning process, and few industrial processes can tolerate the large number of "failures" needed to adapt the controller.

From a practical perspective, one would prefer to let failures take place in a simulated environment (with a model) rather than in a real plant even if the failures are not disastrous or do not cause substantial losses. As opposed to the previous case, this class of neuro-control design is referred to as "model-based neuro-control design". Similar to Equation (3.46), as a result, the problem formulation becomes

$$\text{NN: } \min_{w} F \{y^* - y_m (u, \dots)\} \quad , \quad u = \aleph (w, \dots) \qquad (3.47)$$

Here, y_p in Equation (3.46) is replaced by y_m-the model's output. In this case, knowledge about the processes of interest is required. As can be seen in Figure 3.17, a model replaces the plant/process in the control system.

If a process model is not available, one can first train a second neural network to model the plant dynamics. In the course of modelling the plant, the plant must be operated "normally" instead of being driven out of control. After the modelling stage, the model can then be used for control design. If a plant model is already available, a neural network controller can then be developed in a simulation in which failures cannot cause any loss but that of computer time. A neural network controller after extensive training in the simulation can then be installed in the actual control system.

In fact, these "model-based neuro-control design" approaches have not only proven effective in several studies (Troudet, 1991), but also have already produced notable economic benefits (Staib, 1993). Nevertheless, the quality of control achieved with this approach depends crucially on the quality of the process model. If a model is not accurate enough, the trained neuro-controller is unlikely to perform satisfactorily on the real process. Without an on-line adaptive component, this neuro-controller does not allow for plant drifts or other factors that could adversely affect the performance of the control system.

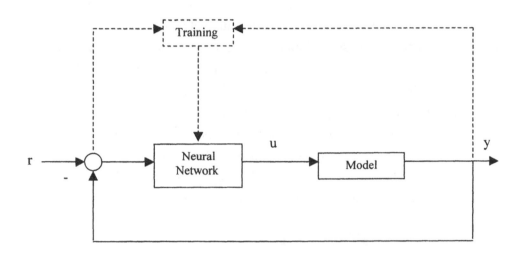

Figure 3.17 A model replaces the plant/process in the control system during the control design phase

The neuro-controller approaches discussed above still share a common shortcoming: A neural network must be trained for every new application. Network retraining is needed even with small changes in the control criterion, such as changes in the relative weighting of control energy and tracking response, or if the controller is to be applied to a different but similar processes. In order to avoid such drawbacks, the concept of "robustness" is naturally brought into the design of a neuro-controller. In robust model-based neuro-control design, a family of process models is considered instead of just a nominal one (see Figure 3.18). Often such a family is specified by a range of noise models or range of the process parameters. Robust neuro-control design can be formulated as follows:

$$\text{NN: } \min_{w} F \{y^* - y_{mi} (u, \dots)\} \; , \; u = \aleph (w, \dots), \quad \forall \, m_i \in \mathbf{M} \qquad (3.48)$$

where m_i stands for the ith member of the model family $\mathbf{M}$. Ideally, the real process to be controlled should belong to this family as well so that the controller is robust not only for the model but also for the real process.

Two aspects of robustness are commonly distinguished. Robust Stability refers to a control system that is stable (qualitatively) over the entire family of processes, whereas robust performance refers to (quantitative) performance criteria being satisfied over the family (Morari & Zafiriou, 1989). Not surprisingly, there is a tradeoff to achieve robustness. By optimizing a neural network controller based upon a fixed (and accurate) process model, high performance can be achieved as long as the process remains invariant, but at the likely cost of brittleness. A robust design procedure, on the other hand, is not likely to achieve the same level of nominal performance but will be less sensitive to process drifts, disturbances, and other sources of process-model mismatch.

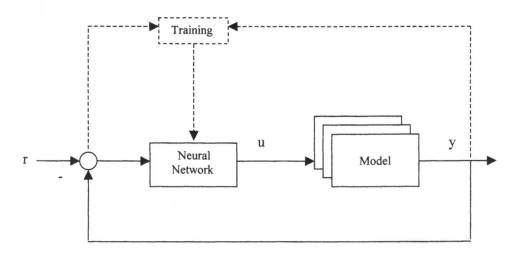

Figure 3.18 Robust model-based neuro-control

3.4.3 Parameterized Neuro-Control

All the above neuro-control approaches share a common shortcoming-the need for extensive application-specific development efforts. Each application requires the optimization of the neural network controller and may also require process model identification. The expense in time and computation has been a significant barrier to widespread implementation of neuro-control systems.

In an attempt to avoid application-specific development, a new neuro-control design concept-parameterized neuro-control (PNC)-has evolved (Samad & Foslien, 1994). Figure 3.19 illustrates this PNC strategy. The PNC controller is equipped with parameters that specify process characteristics and those that provide performance criterion information. For illustration purposes, a PNC can be conceptually formulated as follows:

$$\text{NN: } \min_{w} F(\varepsilon) \{y^* - y_{mi} (\theta, u, \dots)\} \ , \ u = \aleph (w, \hat{\theta}, \varepsilon, \dots) \ , \ \forall \ m_i(\theta) \in \mathbf{M}(\theta)$$

$$(3.49)$$

where ε designates the parameter set that defines the space of performance criteria, θ stands for the process parameter set, $\hat{\theta}$ is the estimates for process parameters, and again $\mathbf{M}(\theta)$ is a family of parameterized models $m_i(\theta)$ in order to account for errors in process parameters estimates θ.

In fact, the two additional types of parameters (ε and θ) make a PNC generic. A PNC is generic in two respects: 1) the process model parameters θ facilitate its application to different processes and 2) the performance parameters ε allow its performance characteristics to be adjustable, or "tunable". For example, if a PNC is designed for first-order plus delay processes, the process parameters (i.e., process gain, time constant, and dead time) will be adjustable parameters to this PNC. Once developed, this PNC requires no application-specific training or adaptation when applied to a first-order plus delay process. It only requires estimates of these process parameters. These estimates do not have to be accurate because the robustness against such inaccuracy is considered in the design phase.

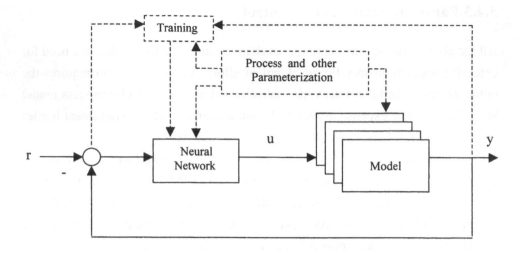

Figure 3.19 Parameterized Neuro-Control

3.5 Summary

In this chapter, we have presented the main ideas underlying Neural Networks and the application of this powerful computational theory to general control problems. We have discussed in some detail the backpropagation learning algorithm for feedforward networks, the integration of fuzzy logic techniques to neural networks to form powerful adaptive neuro-fuzzy inference systems and the basic concepts and current methods of neuro-fuzzy control. At the end, we also gave some remarks about adaptive neuro-control and model-based control of non-linear dynamical systems. In the following chapters, we will show how neural network techniques (in conjunction with other techniques) can be applied to solve real world complex problems of control. This chapter will serve as a basis for the new hybrid intelligent control methods that will be described in Chapter 7 of this book.

Chapter 4

Genetic Algorithms and Fractal Theory for Modelling and Simulation

This chapter introduces the basic concepts and notation of genetic algorithms and simulated annealing, which are two basic search methodologies that can be used for modelling and simulation of complex non-linear dynamical systems. Since both techniques can be considered as general purpose optimization methodologies, we can use them to find the mathematical model which minimizes the fitting errors for a specific problem. On the other hand, we can also use any of these techniques for simulation if we exploit their efficient search capabilities to find the appropriate parameter values for a specific mathematical model. We also present in this chapter the basic concepts and notation of Dynamical Systems and Fractal theory, which are two powerful mathematical theories that enable the understanding of complex non-linear phenomena. Dynamical Systems theory gives us the general framework for treating non-linear systems and enables the identification of the different dynamical behaviors that can occur for a particular dynamic system. On the other hand, Fractal theory gives us powerful concepts and techniques that can be used to measure the complexity of geometrical objects. In particular, the concept of the fractal dimension can be used to measure the geometrical complexity of a data set and this information could be used for modelling as will be illustrated in Chapter 5 of this book.

Genetic algorithms and simulated annealing have been used extensively for both continuous and discrete optimization problems (Jang, Sun & Mizutani, 1997). Common characteristics shared by these methods are described next.

- Derivative freeness: These methods do not need functional derivative information to search for a set of parameters that minimize (or maximize) a given objective function. Instead they rely exclusively on repeated evaluations of the objective function, and the subsequent search direction after each evaluation follows certain heuristic guidelines.

- Heuristic guidelines: The guidelines followed by these search procedures are usually based on simple intuitive concepts. Some of these concepts are motivated by so-called nature's wisdom, such as evolution and thermodynamics.

- Flexibility: Derivative freeness also relieves the requirement for differentiable objective functions, so we can use as complex an objective function as a specific application might need, without sacrificing too much in extra coding and computation time. In some cases, an objective function can even include the structure of a data-fitting model itself, which may be a fuzzy model.

- Randomness: These methods are stochastic, which means that they use random number generators in determining subsequent search directions. This element of randomness usually gives rise to the optimistic view that these methods are "global optimizers" that will find a global optimum given enough computing time. In theory, their random nature does make the probability of finding an optimal solution nonzero over a fixed amount of computation time. In practice, however, it might take a considerable amount of computation time.

- Analytic opacity: It is difficult to do analytic studies of these methods, in part because of their randomness and problem-specific nature. Therefore, most of our knowledge about them is based on empirical studies.

- Iterative nature: These techniques are iterative in nature and we need certain stopping criteria to determine when to terminate the optimization process. Let K denote an iteration count and f_k denote the best objective function obtained at count k; common stopping criteria for a maximization problem include the following:
  1) Computation time: a designated amount of computation time, or number of function evaluations and/or iteration counts is reached.
  2) Optimization goal: f_k is less than a certain preset goal value.
  3) Minimal improvement: $f_k - f_{k-1}$ is less than a preset value.
  4) Minimal relative improvement: $(f_k - f_{k-1})/ f_{k-1}$ is less than a preset value.

Both genetic algorithms (GAs) and simulated annealing (SA) have been receiving increasing amounts of attention due to their versatile optimization capabilities for both continuous and discrete optimization problems. Moreover, both of them are motivated by so-called "nature's wisdom": GAs are based on the concepts of natural selection and evolution; while SA originated in annealing processes found in thermodynamics and metallurgy.

4.1 Genetic Algorithms

Genetic algorithms (GAs) are derivative-free optimization methods based on the concepts of natural selection and evolutionary processes (Goldberg, 1989). They were first proposed and investigated by John Holland at the University of Michigan (Holland, 1975). As a general-purpose optimization tool, GAs are moving out of academia and finding significant applications in many areas. Their popularity can be attributed to their freedom from dependence on functional derivatives and their incorporation of the following characteristics:

- GAs are parallel-search procedures that can be implemented on parallel processing machines for massively speeding up their operations.
- GAs are applicable to both continuous and discrete (combinatorial) optimization problems.
- GAs are stochastic and less likely to get trapped in local minima, which inevitably are present in any optimization application.
- GAs' flexibility facilitates both structure and parameter identification in complex models such as fuzzy inference systems or neural networks.

GAs encode each point in a parameter (or solution) space into a binary bit string called a "chromosome", and each point is associated with a "fitness value" that, for maximization, is usually equal to the objective function evaluated at the point. Instead of a single point, GAs usually keep a set of points as a "population", which is then evolved repeatedly toward a better overall fitness value. In each generation, the GA constructs a new population using "genetic operators" such as crossover and mutation; members with higher fitness values are more likely to survive and to participate in mating (crossover) operations. After a number of generations, the population contains members with better fitness values; this is analogous to Darwinian models of evolution by random mutation and natural selection. GAs and their variants are sometimes referred to as methods of "population-based optimization" that improve performance by upgrading entire populations rather than individual members. Major components of GAs include encoding schemes, fitness evaluations, parent selection, crossover operators, and mutation operators; these are explained next.

Encoding schemes: These transform points in parameter space into bit string representations. For instance, a point (11, 4, 8) in a three-dimensional parameter space can be represented as a concatenated binary string:

$$\underbrace{1011}_{11} \quad \underbrace{0100}_{4} \quad \underbrace{1000}_{8}$$

in which each coordinate value is encoded as a "gene" composed of four binary bits using binary coding. other encoding schemes, such as gray coding, can also be used and, when necessary, arrangements can be made for encoding negative, floating-point, or discrete-valued numbers. Encoding schemes provide a way of translating problem-specific knowledge directly into the GA framework, and thus play a key role in determining GAs' performance. Moreover, genetic operators, such as crossover and mutation, can and should be designed along with the encoding scheme used for a specific application.

Fitness evaluation: The first step after creating a generation is to calculate the fitness value of each member in the population. For a maximization problem, the fitness value f_i of the ith member is usually the objective function evaluated at this member (or point). We usually need fitness values that are positive, so some kind of monotonical scaling and/or translation may by necessary if the objective function is not strictly positive. Another approach is to use the rankings of members in a population as their fitness values. The advantage of this is that the objective function does not need to be accurate, as long as it can provide the correct ranking information.

Selection: After evaluation, we have to create a new population from the current generation. The selection operation determines which parents participate in producing offspring for the next generation, and it is analogous to "survival of the fittest" in natural selection. Usually members are selected for mating with a selection probability proportional to their fitness values. The most common way to implement this is to set the selection probability equal to:

$$f_i / \sum_{k=1}^{k=n} f_k,$$

where n is the population size. The effect of this selection method is to allow members with above-average fitness values to reproduce and replace members with below-average fitness values.

Crossover: To exploit the potential of the current population, we use "crossover" operators to generate new chromosomes that we hope will retain good features from the previous generation. Crossover is usually applied to selected pairs of parents with a probability equal to a given "crossover rate". "One-point crossover" is the most basic crossover operator, where a crossover point on the genetic code is selected at random and two parent chromosomes are interchanged at this point. In "two-point crossover", two crossover points are selected and the part of the chromosome string between these two points is then swapped to generate two children. We can define n-point crossover similarly. In general, (n-1)-point crossover is a special case of n-point crossover. Examples of one-and two-point crossover are shown in Figure 4.1.

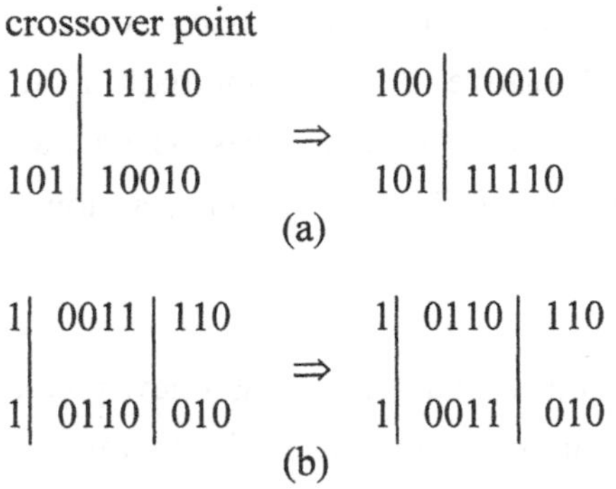

$$
\begin{array}{c}
\text{crossover point} \\
\begin{array}{c|c}
100 & 11110 \\[4pt]
101 & 10010
\end{array}
\qquad \Rightarrow \qquad
\begin{array}{c|c}
100 & 10010 \\[4pt]
101 & 11110
\end{array} \\
\text{(a)}
\end{array}
$$

$$
\begin{array}{c}
\begin{array}{c|c|c}
1 & 0011 & 110 \\[4pt]
1 & 0110 & 010
\end{array}
\qquad \Rightarrow \qquad
\begin{array}{c|c|c}
1 & 0110 & 110 \\[4pt]
1 & 0011 & 010
\end{array} \\
\text{(b)}
\end{array}
$$

Figure 4.1 Crossover operators: (a) one-point crossover; (b) two-point crossover.

Mutation: Crossover exploits current gene potentials, but if the population does not contain all the encoded information needed to solve a particular problem, no amount of gene mixing can produce a satisfactory solution. For this reason, a "mutation" operator capable of spontaneously generating new chromosomes is included. The most common way of implementing mutation is to flip a bit with a probability equal to a very low given "mutation rate". A mutation operator can prevent any single bit from converging to a value throughout the entire population and, more important, it can prevent the population from converging and stagnating at any local optima. The mutation rate is usually kept low so good chromosomes obtained from crossover are not lost. If the mutation rate is high (above 0.1), GA performance will approach that of a primitive random search. Figure 4.2 provides an example of mutation.

Mutated bit
↓
10011110   ⇒   10011010

Figure 4.2 Mutation operator.

In the natural evolutionary process, selection, crossover, and mutation all occur in the single act of generating offspring. Here we distinguish them clearly to facilitate implementation of and experimentation with GAs.

Based on the aforementioned concepts, a simple genetic algorithm for maximization problems is described next.

Step 1: Initialize a population with randomly generated individuals and evaluate the fitness value of each individual.

Step 2: Perform the following operations:

(a) Select two members from the population with probabilities proportional to their fitness values.

(b) Apply crossover with a probability equal to the crossover rate.

(c) Apply mutation with a probability equal to the mutation rate.

(d) Repeat (a) to (d) until enough members are generated to form the next generation.

Step 3: Repeat steps 2 and 3 until a stopping criterion is met.

Figure 4.3 is a schematic diagram illustrating how to produce the next generation from the current one.

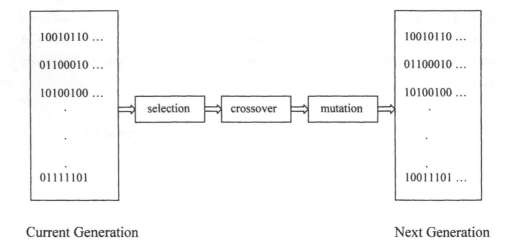

Current Generation   Next Generation

Figure 4.3 Producing the next generation in GAs.

4.2 Simulated Annealing

"Simulated Annealing" (SA) is another derivative-free optimization method that has recently drawn much attention for being as suitable for continuous as for discrete (combinational) optimization problems (Otten & Ginneken, 1989). When SA was first proposed (Kirkpatrick, Gelatt & Vecchi, 1983) it was mostly known for its effectiveness in finding near optimal solutions for large-scale combinatorial optimization problems, such as traveling salesperson problems and placement problems. Recent applications of SA and its variants (Ingber & Rosen, 1992) also demonstrate that this class of optimization approaches can be considered competitive with other approaches when there are continuous optimization problems to be solved.

Simulated annealing was derived from physical characteristics of spin glasses (Kirkpatrick, Gelatt & Vecchi, 1983). The principle behind simulated annealing is analogous to what happens when metals are cooled at a controlled rate. The slowly falling temperature allows the atoms in the molten metal to line themselves up and form a regular crystalline structure that has high density and low energy. But if the temperature goes down too quickly, the atoms do not have time to orient themselves into a regular structure and the result is a more amorphous material with higher energy.

In simulated annealing, the value of an objective function that we want to minimize is analogous to the energy in a thermodynamic system. At high temperatures, SA allows function evaluations at faraway points and it is likely to accept a new point with higher energy. This corresponds to the situation in which high-mobility atoms are trying to orient themselves with other nonlocal atoms and the energy state can occasionally go up. At low temperatures, SA evaluates the objective function only at local points and the likelihood of it accepting a new point with higher energy is much lower. This is analogous to the situation in which the low-mobility atoms can only orient themselves with local atoms and the energy state is not likely to go up again.

Obviously, the most important part of SA is the so-called "annealing schedule" or "cooling schedule", which specifies how rapidly the temperature is lowered from high to low values. This is usually application specific and requires some experimentation by trial-and-error.

Before giving a detailed description of SA, first we shall explain the fundamental terminology of SA.

Objective function: An objective function $f(.)$ maps an input vector x into a scalar E:
$$E = f(x),$$
where each x is viewed as a point in an input space. The task of SA is to sample the input space effectively to find an x that minimizes E.

Generating function: A generating function $g(.,.)$ specifies the probability density function of the difference between the current point and the next point to be visited. Specifically, Δx ($= x_{new} - x$) is a random variable with probability density function $g(\Delta x, T)$, where T is the temperature. For common SA (especially in combinatorial optimization applications), $g(.,.)$ is usually a function independent of the temperature T.

Acceptance function: After a new point x_{new} has been evaluated, SA decides whether to accept or reject it based on the value of an acceptance function $h(.\,,.)$. The most frequently used acceptance function is the "Boltzmann probability distribution":

$$h(\Delta E, T) = \frac{1}{1 + \exp(\Delta E / (cT))} \qquad (4.1)$$

where c is a system-dependent constant, T is the temperature, and ΔE is the energy difference between x_{new} and x:

$$\Delta E = f(x_{new}) - f(x)$$

The common practice is to accept x_{new} with probability $h(\Delta E, T)$.

Annealing schedule: An annealing schedule regulates how rapidly the temperature T goes from high to low values, as a function of time or iteration counts. The exact interpretation of "high" and "low" and the specification of a good annealing schedule require certain problem-specific physical insights and/or trial-and-error. The easiest way of setting an annealing schedule is to decrease the temperature T by a certain percentage at each iteration.

The basic algorithm of simulated annealing is the following:

Step 1: Choose a start point x and set a high starting temperature T. Set the iteration count k to 1.

Step 2: Evaluate the objective function $E = f(x)$.

Step 3: Select Δx with probability determined by the generating function $g(\Delta x, T)$. Set the new point x_{new} equal to $x + \Delta x$.

Step 4: Calculate the new value of the objective function: $E_{new} = f(x_{new})$.

Step 5: Set x to x_{new} and E to E_{new} with probability determined by the acceptance function $h(\Delta E, T)$, where $\Delta E = E_{new} - E$.

Step 6: Reduce the temperature T according to the annealing schedule (usually by simply setting T equal to ηT, where η is a constant between 0 and 1).

Step 7: Increment iteration count k. If k reaches the maximum iteration count, stop the iterating. Otherwise, go back to step 3.

In conventional SA, also known as "Boltzmann machines", the generating function is a Gaussian probability density function:

$$g(\Delta x , T) = (2\pi T)^{-n/2} \exp[-\| \Delta x \|^2 / (2T)] \qquad (4.2)$$

where Δx ($= x_{new} - x$) is the deviation of the new point from the current one, T is the temperature, and n is the dimension of the space under exploration. It has been proven (Geman & Geman, 1984) that a Boltzman machine using the aforementioned generating function $g(. , .)$ can find a global optimum of $f(x)$ if the temperature T is reduced no faster than $T_0 / \ln k$.

Variants of Boltzmann machines include the "Cauchy machine" or "fast simulated annealing" (Szu & Hartley, 1987), where the generating function is the Cauchy distribution:

$$g (\Delta x) = \frac{T}{(\| \Delta x \|^2 + T^2)^{(n+1)/2}} \qquad (4.3)$$

The fatter tail of the Cauchy distribution allows it to explore farther from the current point during the search process.

Another variant of the original SA, the so-called "very fast" simulated annealing (Ingber & Rosen, 1992), was designed for optimization problems in a constrained search space. Very fast simulated annealing has been reported to be faster than genetic algorithms on several test problems by the same authors.

4.3 Basic Concepts of Fractal Theory

In this section we present a brief overview of the field of Non-Linear Dynamical Systems and Fractal Theory. Recently research has shown that many simple non-linear deterministic systems can behave in an apparently unpredictable and "chaotic" manner (Grebogi, Ott, & Yorke, 1987). The existence of complicated dynamics has been discussed in the mathematical literature for many decades with important contributions by Poincaré, Birkhoft, Smale and Kolmogorov and his students, among others. Nevertheless, it is only recently that the wide-ranging impact of "chaos" has been recognized. Consequently, the field is now undergoing explosive growth, and many applications have been made across a broad spectrum of scientific disciplines-robotics, engineering, physics, chemistry, fluid mechanics and economics, to name several. We start with some basic definitions of concepts used in this book.

Dynamic System: This is a set of mathematical equations that allows one, in principle, to predict the future behavior of the system given the past. One example is a system of first-order ordinary differential equations in time:

$$\frac{dx}{dt} = G(x,t) \tag{4.4}$$

where $x(t)$ is a D-dimensional vector and G is a D-dimensional vector function of x and t. Another example is a map.

Map: A map is an equation of the following form:

$$x_{t+1} = F(x_t) \tag{4.5}$$

where the "time" t is discrete and integer valued. Thus, given x_0, the maps gives x_1. Given x_1, the map gives x_2, and so on.

Dissipative system: In Hamiltonian (conservative) systems such as the ones arising in Newtonian mechanics of particles (without friction), phase space volumes are preserved by time evolution (the phase space is the space of variables that specify the state of the system). Consider, for example, a two-dimensional phase space (q, p), where q denotes a position variable and p a momentum variable. Hamilton's equations of motion take the set of initial conditions at time t $=t_0$ and evolve them in time to the set at time $t = t_1$. Although the shapes of the sets are different, their areas are the same. By a dissipative system we mean one that does not have this property. Areas should typically decrease (dissipate) in time so that the area of the final set would be less than the area of the initial set. As a consequence of this, dissipative systems typically are characterized by the presence of attractors.

Attractor: If one considers a system and its phase space, then the initial conditions may be attracted to some subset of the phase space (the attractor) as time t $\rightarrow \infty$. For example, for a damped harmonic oscillator the attractor is the point at rest.

For a periodically driven oscillator in its limit cycle the limit set is a closed curve in the phase space.

Strange attractor: In the above two examples, the attractors were a point, which is a set of dimension zero, and closed curve, which is a set of dimension one. For many other attractors the attracting set can be much more irregular (some would say pathological) and, in fact, can have a dimension that is not an integer. Such sets have been called "fractal" and, when they are attractors, they are called strange attractors. The existence of a strange attractor in a physically interesting model was first demonstrated by Lorenz (see Lorenz, 1963).

Chaotic attractor: By this term we mean that if we take two typical points on the attractor that are separated from each other by a small distance $\Delta(0)$ at t = 0, then for increasing t they move apart exponentially fast. That is, in some average sense:

$$\Delta(t) \sim \Delta(0) \exp(\lambda t) \qquad (4.6)$$

with $\lambda > 0$ (where λ is called the Lyapunov exponent). Thus a small uncertainty in the initial state of the system rapidly leads to inability to forecast its future. It is typically the case that strange attractors are also chaotic.

One of the most prominent, chaotic, continuous-time dynamical systems is the "Lorenz attractor", named after the meteorologist E.N. Lorenz who investigated the three-dimensional, continuous-time system

$$x' = S(-x + y)$$
$$y' = rx - y - xz \qquad s, r, b > 0 \qquad (4.7)$$
$$z' = -bz + xy$$

emerging in the study of turbulence in fluids. For r above the critical value of r = 28.0, trajectories of Equation (4.7) evolve in a rather unexpected way. Suppose that a trajectory starts at an initial value near the origin. For some time the trajectory regularly spirals outward from one fixed point, then the trajectory jumps to a region near another fixed point and does the same thing. As trajectories starting at different initial values all converge to and remain in the same region near the two fixed points, the region is considered an "attractor". It is a "strange attractor" because it is neither a point nor a closed curve. In general, this chaotic behavior can only occur for systems of at least three simultaneous non-linear differential equations or for systems of at least a one-dimensional non-linear map (Devaney, 1989).

Fractal geometry is a mathematical tool for dealing with complex systems that have no characteristic length scale. A well known example is the shape of a coastline. When we see two pictures of a coastline on two different scales, we cannot tell which scale belongs to which picture: both look the same. This means that the coastline is scale invariant or, equivalently, has no characteristic length scale. Other examples in nature are rivers, cracks, mountains, and clouds. Scale-invariant systems are usually characterized by noninteger ("fractal") dimensions.

The dimension tell us how some property of an object or space changes as we view it at increased detail. There are several different types of dimension. The fractal dimension d_f describes the space filling properties of an object. Three examples of the fractal dimension are the self-similarity dimension, the capacity dimension, and the Hausdorff-Besicovitch dimension. The topological dimension d_T describes how points within an object are connected together. The embedding dimension d_e describes the space in which the object is contained.

The fractal dimensions d_f are useful and important tools to quantify self-similarity and scaling. Essentially, the dimension tell us how many new pieces are resolved as the resolution is increased. The self-similarity dimension can only be applied to geometrical self-similar objects, where the small pieces are exact copies of the whole object. However the capacity dimension can be used to analyze irregularly shaped objects that are statistically self-similar. On the other hand, the Hausdorff-Besicovitch dimension requires more complex mathematical tools. For this reason, we will limit our discussion here to the capacity dimension.

A ball is the set of points within radius r of a given point. We determine N(r) the minimum number of balls required so that each point in the object is contained within at least one ball of radius r. In order to cover all the points of the object, the balls may need to overlap. The capacity dimension is defined by the following equation:

$$d_c = \lim_{r \to 0} \frac{\log N(r)}{\log(1/r)} . \tag{4.8}$$

The capacity dimension defined as above is a measure of the space filling properties of an object because it gives us an idea of how much work is needed to cover the object with balls of changing size.

A useful method to determine the capacity dimension is to choose balls that are the non-overlapping boxes of a rectangular coordinate grid. $N(r)$ is then the number of boxes with side of length r that contain at least one point of the object. Efficient algorithms have been developed to perform this "box counting" for different values of r, and thus determine the box counting dimension as the best fit of log $N(r)$ versus $\log(1/r)$.

The fractal dimension d_f characterizes the space-filling properties of an object. The topological dimension d_T characterizes how the points that make up the object are connected together. It can have only integer values. Consider a line that is so long and wiggly that it touches every point in a plane and thus covers an area. Because it covers a plane, its space-filling fractal dimension $d_f = 2$. However, no matter how wiggly it is, it is still a line and thus has topological dimension $d_T=1$. Thus, the essence of a fractal is that its space-filling properties are larger than one anticipates from its topological dimension. Thus we can now present a formal definition of a fractal (Mandelbrot, 1987), namely, that an object is a fractal if and only if $d_f > d_T$.
However, there is no one definition that includes all the objects or processes that have fractal properties.

Despite the identification of fractals in nearly every branch of science, too frequently the recognition of fractal structure is not accompanied with any additional insight as to its cause. Often we do not even have the foggiest idea as to the underlying dynamics leading to the fractal structure. The chaotic dynamics of non-linear systems, on the other hand, is one area where considerable progress has been made in understanding the connection with fractal geometry. Indeed, chaotic dynamics and fractal geometry have such a close relationship that one of the hallmarks of chaotic behavior has been the manifestation of fractal geometry, particularly for strange attractors in dissipative systems (Rasband, 1990). For a practical definition we take a "strange attractor", for a dynamic system, to be an attracting set with fractal dimension. For example, the famous Lorenz strange attractor has a fractal dimension of about 2.06. Also, we think that beyond only this relationship between strange attractors and the fractal dimension of the set, there is a deeper relationship between the underlying dynamics of a system and the fractal nature of its behavior. We will explore this relationship in more detail in the following chapter.

4.4 Summary

In this chapter, we have presented the main ideas underlying Dynamical Systems and Fractal theory and we have only started to point out the many possible applications of these two powerful mathematical theories. We have discussed in some detail the concepts of strange attractors, chaotic behavior and fractal dimension. The concept of the fractal dimension will be the basis of the method for time series analysis that will be used in Chapter 5 to achieve Automated Mathematical Modelling of dynamic systems. Also, we have introduced two basic intelligent search methodologies that can be used for mathematical modelling and simulation. We have described in some detail how genetic algorithms can be used for the optimization of non-linear functions. Genetic algorithms can be used for modelling by defining an appropriate objective function or they can be used for simulation if they are aimed mainly at searching the parameter space (of the models) in an efficient way. In Chapter 6, we will explore this approach to achieve automated simulation of non-linear dynamical systems. We have also described in this chapter an alternative search method called simulated annealing, which is also a good choice for optimization problems.

Chapter 5

Fuzzy-Fractal Approach for Automated Mathematical Modelling

We describe in this chapter a new method to perform automated mathematical modelling for non-linear dynamic systems using SC techniques, Dynamical Systems Theory and Fractal Theory. The idea of using Dynamical Systems Theory (DST) and Fractal Theory (FT) as alternative approaches for modelling can be justified if we consider that traditional statistical methods only have limited success in real world complex applications, and this is mainly because many real problems show very complicated dynamics in time. Traditional statistical methods assume that the erratic behavior of a time series is mainly due to a external random error (that can not be explained (Castillo & Melin, 1994)). However, a DST approach, using non-linear mathematical models, can explain this erratic behavior because "chaos" is an intrinsic part of this type of models (Castillo & Melin, 1995a). It is a well known fact from DST (see Devaney, 1989), that even very simple non-linear mathematical models can exhibit the behavior known as "chaos" for certain parameter values, and therefore are good candidates to use as equations for modelling complex dynamic systems (Castillo & Melin, 1995b). Fractal Theory (see Mandelbrot, 1987), also offers a way to explain the erratic behavior of a time series, but the method is geometrical in the sense that the fractal dimension is used to describe the complexity of the distribution of the data points (Castillo & Melin, 1996a).

We describe a prototype implementation of our new method for Automated Mathematical Modelling (AMM) as a computer program written in the PROLOG programming language (Bratko, 1990). This computer program can be considered an intelligent system for the domain of Non-linear Dynamical Systems (NDS) because it uses SC techniques to obtain the "best" mathematical model for a given dynamic system. The use of SC techniques is to achieve the goal of automated modelling of NDS by simulating (in the computer) how human experts in this domain obtain the "best" model for a given problem. Given a specific time series the intelligent system develops mathematical models based on the geometry of the data. The method for AMM consists of three main parts: Time Series Analysis, Developing a Set of Admissible Models and Selecting the Best Model. First, the intelligent system uses the fractal dimension to classify the components of the time series over a set of qualitative values, then the system uses this information to decide (using a fuzzy rule base) which dynamical models are the most appropriate for the data, and finally the system decides which model is the "best" one using heuristics and statistical calculations. The use of Fuzzy logic in real-world applications has been now well recognized and many systems have been developed (Yamamoto & Yun, 1997). In this case, we came to the conclusion that the best way to convey the information of modelling problems was using fuzzy sets (Badiru, 1992). Also, we think that the best way to reason with uncertainty in this case is using Fuzzy Logic.

The intelligent system develops only the kind of mathematical models that are more likely to give a "good" prediction based on the knowledge that human experts have about this matter. This knowledge is contained in the knowledge base of the intelligent system, and is the main factor in limiting the number of models that the system explores (Castillo & Melin, 1996b). The intelligent system also has some generalized knowledge about the mathematical models that we expect to discover in the NDS domain (Castillo & Melin, 1996c). This knowledge is expressed as families of parameterized mathematical models.

5.1 The Problem of Automated Mathematical Modelling

The problem of achieving automated mathematical modelling can be defined formally as follows (Castillo & Melin, 1994) :

Given: A data set (time series) with m data points, $D = \{d_1, d_2, ..., dm\}$ where $di \in R^n$, $i = 1,...,m$, and $n = 1, 2,...$.

Goal: From the data set D, discover automatically the "best" mathematical model for the time series.

This problem is not a simple one, because in theory there can be an infinite number of mathematical models that can be build for a given data set (Rao & Lu, 1993). So the problem lies in knowing which models to try for a data set and then to select the "best" one. Let M be the space of mathematical models defined for a given data set D. Let $MA = \{M_1, ..., Mq\}$ be the set of admissible models that are considered to be appropriate for the geometry of the data set D. The problem is to find automatically the "best" model Mb for time series prediction (Castillo & Melin, 1995a).

We can consider mathematical statistical models of the following form:
$$Y = F(X) + \varepsilon (0,\sigma) \tag{5.1}$$
where $\varepsilon (0,\sigma)$ represents a 0-mean Gaussian noise-process with standard deviation σ. F(X) is a polynomial equation in X, where the p predictor variables are in the vector:
$$X = (X_1, X_2, ..., Xp).$$
We can also consider mathematical models as "dynamical systems" of the following form:
$$dY/dt = F(Y) \tag{5.2}$$
where Y is a vector of variables of the form (p is the number of variables):
$$Y = (Y_1, Y_2, ..., Yp)$$
and F(Y) is a non-linear function of Y. Other kind of mathematical models are the discrete "dynamical systems" of the following form:
$$Y_t = F(X) \tag{5.3}$$

where $X = (Y_{t-1}, Y_{t-2}, ..., Y_{t-p})$ and $F(X)$ is a non-linear function of X. Note that in these last two cases we have deterministic models expressed as differential or difference equations.

We consider the use of the fractal dimension as a mathematical model of the time series in the following form:

$$d = [\log(N)/\log(1/r)] \tag{5.4}$$

where d is the fractal dimension for an object of N parts, each scaled down by a ratio r. For an estimation of this dimension we can use the following equation:

$$N(r) = \beta[\ 1/r^d\ ] \tag{5.5}$$

where $N(r)$ = number of boxes contained in a geometrical object and r = size of the box. We can obtain the box dimension of a geometrical object (Mandelbrot, 1987) counting the number of boxes for different sizes and performing a logarithmic regression on this data. For our particular case the geometrical object consists of the curve constructed using the set of points from the time series. We show in Figure 5.1 (a) the curve and the boxes used to cover it. In Figure 5.1 (b) the corresponding logarithmic regression is illustrated.

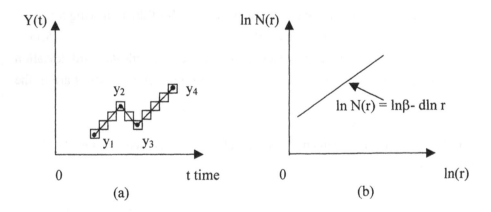

Figure 5.1 Fractal dimension of a time series: (a) the curve and the boxes covering it, (b) the logarithmic regression to find d

The models for the statistical methods can be linear as well as non-linear equations. We show below some sample statistical models (Gujarati, 1987) that can be used for mathematical modelling:

a) linear regression: $Y_t = a + bt$ (5.6)

b) quadratic regression: $Y_t = a + bt + ct^2$ (5.7)

c) logarithmic regression: $\ln Y_t = a + b\ln t$ (5.8)

d) semi-log regression: $Y_t = a + b\ln t$ (5.9)

e) first order autoregression: $Y_t = a + bY_{t-1}$ (5.10)

f) second order autoregression: $Y_t = a + bY_{t-1} + cY_{t-2}$ (5.11)

The mathematical models for continuous dynamical systems can be one-dimensional, two-dimensional, three-dimensional and so on. We show below some sample models (Rasband, 1990) that can be used for mathematical modelling:

a) Logistic differential equation:

$$dY_1/dt = a\, Y_1(1 - Y_1)$$ (5.12)

b) Lotka Volterra two dimensional:

$$dY_1/dt = aY_1 - bY_1Y_2$$
$$dY_2/dt = bY_1Y_2 - cY_2$$ (5.13)

c) Lotka Volterra three dimensional:

$$dY_1/dt = Y_1(1 - Y_1 - aY_2 - bY_3)$$
$$dY_2/dt = Y_2(1 - bY_1 - Y_2 - aY_3)$$ (5.14)
$$dY_3/dt = Y_3(1 - aY_1 - bY_2 - Y_3)$$

d) Lorenz three dimensional:

$$dY_1/dt = aY_2 - aY_1$$
$$dY_2/dt = -Y_1Y_3 + bY_1 - Y_2$$ (5.15)
$$dY_3/dt = Y_1Y_2 - cY_3$$

The mathematical models for discrete dynamical systems can also be one, two, three dimensional or more. We show below some sample models (Rasband, 1990) that can be used for mathematical modelling:

a) Logistic difference equation:

$$Y_{t+1} = aY_t(1 - Y_t)$$ (5.16)

b) Logistic two dimensional difference equation:

$$Y_{t+1} = X_t \tag{5.17}$$
$$X_{t+1} = aX_t(1 - X_t)$$

c) Lotka Volterra two dimensional:

$$Y_{t+1} = aY_t - bY_tX_t \tag{5.18}$$
$$X_{t+1} = bY_tX_t - cX_t$$

d) Henon map two dimensional:

$$Y_{t+1} = X_t \tag{5.19}$$
$$X_{t+1} = a - X_t^2 + bY_t$$

In all of the above mathematical models a, b and c are parameters that need to be estimated using the corresponding numerical methods. For example, for the regression models we can use the least squares method (Gujarati, 1987) for parameter estimation, but for the differential equations we need to use a Gauss-Newton type method (Jang, Sun & Mizutani, 1997).

5.2 A Fuzzy-Fractal Method for Automated Modelling

In this section, we show how SC techniques can be used to automate the process of discovering the best model for a given dynamical system. The human experts usually try several (in some cases many) mathematical models before they are satisfied with the "goodness" of one model. The experts use their knowledge about modelling problems in a specific domain to limit their search of models, in this way obtaining the "best" results as quickly as possible. The main goal of our work was to capture this knowledge of modelling in a computer program, in this way obtaining a software tool capable of emulating intelligent behavior in the domain of NDS (Castillo & Melin, 1997a). In the remaining of this section we describe the basic algorithm for discovering mathematical models, then in the following section the implementation of the algorithm as a computer program in the PROLOG programming language.

Our new method for solving the automated modelling problem is based on several novel ideas. We consider that the modelling problem can be divided in three main parts: Time Series Analysis, Selection of Appropriate Models and Selection of the Best Model. The first part of the problem consists in obtaining the time series components from the data. Our solution to this part of the problem is a new classification scheme based on the notion of the fractal dimension . This classification scheme is a one to one map between the fractal dimension of the data set and the qualitative values for the components of the time series (Castillo & Melin, 1995a). Once this part of the problem is solved, the second part consists in simulating an expert decision process that gives us the set of Mathematical Models appropriate for the geometry of the time series. This expert decision process is simulated using SC techniques and is the main part of the method for AMM (Castillo & Melin, 1997b). The third part of the problem consists in designing a method to compare all the models obtained in the second part, to obtain which one is the "best" model for the given time series. Our method to compare all the models has to consider statistical measures of goodness between non-linear dynamical systems and linear statistical models, to decide which model fits best the data set (time series).

The new algorithm for automated mathematical modelling is shown in Figure 5.2.

STEP 1   Read the data set $D = \{d_1, d_2, ..., dm\}$.

STEP 2   Time Series Analysis of the set D to find the components.

STEP 3   Find the set of Admissible models $MA = \{M_1, M_2, ..., Mq\}$, using the qualitative values of the time series components.

STEP 4   Find the "Best" mathematical model Mb from the set MA using the measures of "goodness" of each of the models from the set MA.

Figure 5.2 Algorithm for automated mathematical modelling

We call this algorithm IDIMM (for Intelligent Discovery of Mathematical Models) and is an integration of SC techniques with Dynamical Systems Theory, Fractal Theory and Statistical Methods, to obtain mathematical models for a given time series. In the following section we will show how this algorithm can be implemented to achieve the goal of AMM for Dynamical Systems.

5.3 Implementation of the Method for Automated Modelling

The implementation of the new method for AMM as a computer program was done using the PROLOG programming language (the complete program for general dynamical systems is shown in Appendix A). The choice of PROLOG is because of its symbolic manipulation features and also because it is an excellent language for developing Prototypes (Bratko, 1990). The computer program was developed using an architecture very similar to that of an intelligent system (knowledge base, inference engine and user interface) with the addition of a numerical module for parameter estimation (see Figure 5.3). We will focus our description of implementation details only to the knowledge base of the intelligent system because this is the most important part of the computer program. In the program, the knowledge base is the part that simulates the process of model discovery described by steps 2 to 4 in the IDIMM algorithm of the last section. Accordingly, the knowledge base consists of three Expert Modules: Time Series Analysis, Expert Selection and Best Model Selection. In the following lines we will describe each of these modules.

5.3.1 Description of the Time Series Analysis Module

This module is the implementation of Step 2 of the IDIMM algorithm and contains the knowledge necessary for time series analysis, i.e., the knowledge to extract from the data the time series components. Our method for time series analysis consist in the use of the fractal dimension of the set of points D as a measure of the geometrical complexity of the time series (Castillo & Melin,

1996b). We use the value of the fractal dimension to classify the time series components over a set of qualitative values. Our classification scheme was obtained by a combination of expert knowledge and mathematical modelling for several samples of data. To give an idea of this scheme we show in Table 5.1 some sample rules of this module.

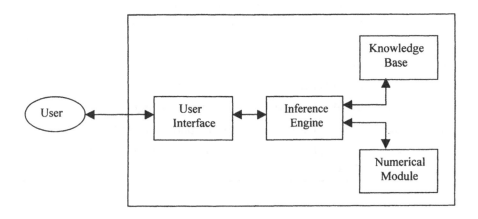

Figure 5.3 General architecture of the intelligent system

The basic idea behind the rule base of Table 5.1 is that when the fractal dimension is close to one, we have a data set resembling a line, and when the fractal dimension is near two, we have a data set with very rapid oscillations (almost covering a finite area). We performed several experiments with real data sets to decide on the classification needed for this "Time Series Analysis Module" and we found that for the moment classifying the periodic components in "simple", "regular", "difficult" and "chaotic" was sufficient. Also, we only classify the "trend" component in two kinds: "linear" and "non-linear". Of course, it is possible that we may need a better classification in the future, for a more accurate implementation of this Module, but now we are only showing how the method can be implemented.

Table 5.1 Sample rules for time series analysis

IF	THEN
Fractal_dimension(D)$\in$(0.8,1.2)	Trend = linear, Time_series = smooth
Fractal_dimension(D)$\in$[1.2,1.5)	Trend = non_linear, Time_series = cyclic
Fractal_dimension(D)$\in$[1.5,1.8)	Time_series = erratic
Fractal_dimension(D)$\in$[1.2,1.4)	Periodic_part = simple
Fractal_dimension(D)$\in$[1.4,1.6)	Periodic_part = regular
Fractal_dimension(D)$\in$[1.6,1.7)	Periodic_part = difficult
Fractal_dimension(D)$\in$[1.7,1.8)	Periodic_part = very_difficult
Fractal_dimension(D)$\in$[1.8,2]	Periodic_part = chaotic

In conclusion, our method for time series analysis consists of a one to one mapping between the fractal dimension of the set D and the qualitative values of the time series components. This set of qualitative values for the components is the information needed as input for the "Expert Selection Module" (implementing Step 3 of the algorithm) which will be described next.

5.3.2 Description of the Expert Selection Module

This module is the implementation of the step 3 of the IDIMM algorithm and contains the knowledge necessary to select the kind of mathematical models more appropriate for the type of data given, i.e., given the qualitative values of the time series components decide which models are more likely to give a good prediction. Our method for selecting the models consists of a set of fuzzy rules (heuristics) that simulates the human expert decision process of model selection (Castillo &

Melin, 1997b). In our approach the qualitative values of the time series components are viewed as fuzzy sets (using the fractal dimension as a classification variable). We have membership functions for each of the qualitative values of the time series components. Also, the qualitative values of the "Type_Model" variable are considered as fuzzy sets and we have membership functions for each of these values. To give and idea of the way this Expert knowledge is structured, we show in Table 5.2 some sample rules of this module (Castillo & Melin, 1998a).

The rules in Table 5.2 show how this expert module selects the appropriate models for a given dynamical system, using as information the dimensionality of the problem (number of variables, which are the "Dim" values in Table 5.2) and the qualitative values of the time series components. Each rule of this table contains a piece of knowledge about the problem of model selection.

We have to mention here that the role of Fuzzy Logic is very important for this module, because it enables the simulation of the expert reasoning process under uncertainty. We came to the conclusion that the rules, for deciding which models are appropriate for a given time series, can't be categorical because the complexity of the modelling problems is very high. Since, it is well known that fuzzy logic has been applied successfully to problems in Engineering and Manufacturing (Badiru, 1992), and our modelling problem required reasoning under uncertainty, we decided to use fuzzy logic techniques. In the following lines we will explain how the knowledge of the experts is contained in the fuzzy rules of this module with an example.

Suppose that a Time Series Analysis on a particular data set (time series) for a one-dimensional problem results in a Trend component valued as "non-linear" with a fractal dimension of 1.37, and a Periodic component valued as "simple" with the same fractal dimension, then the logical conclusion is that the "Logistic Map" is the best model for this problem with a 90% degree of certainty. Of course, other mathematical models have a lower degree of certainty for this particular example. The reasoning behind this rule is that a time series that exhibits a non-linear trend and simple periodicity can be modeled by a logistic map with relatively good accuracy.

Table 5.2 Sample fuzzy rules for model selection

IF			THEN
Dim	Trend	Periodic_part	Type_Model
one	non_linear	simple	logistic_differential_equation
two	non_linear	simple	lotka_volterra_differential_equation
three	non_linear	regular	lorenz_differential_equation
one	non_linear	simple	logistic_difference_equation
two	non_linear	regular	lotka_volterra_difference_equation

5.3.3 Description of the Best Model Selection Module

This module is the implementation of step 4 of the IDIMM algorithm and contains the knowledge to select the "best" mathematical model for prediction, i.e., given the set of selected models generated by step 3, decide which model is the "best" one to predict the time series. Our method for selecting the "best" model consists of comparing the Sum of Squares of Errors (SSE) for all the models and selecting the one that minimizes SSE. This criteria has the advantage of been valid for all the types of models that we consider for the intelligent system (statistical models and non-linear dynamical models). The reasoning behind this criteria is that the value of the SSE is a measure of how well a particular mathematical model fits the data (time series) for a given problem. To give an idea of our method, we show in Table 5.3 a sample case where the set of selected models is:

$$MS = \{M_1, M_2, M_3, M_4, M_5, M_6\}$$

and the model with the lowest SSE is M_4.

Table 5.3 Method for best model selection using the SSE

MODEL		TYPE	SSE	BEST MODEL
M_1:	$Y = a_1 + b_1 t$	Statistical	SSE_1	no
M_2:	$Y = a_2 + b_2 t + c_2 t^2$	Statistical	SSE_2	no
M_3:	$\ln Y = \ln a_3 + b_3 \ln t$	Statistical	SSE_3	no
M_4:	logistic_differential_eq	Dynamical	SSE_4	yes
M_5:	lotka_volterra_difference_eq	Dynamical	SSE_5	no
M_6:	lorenz_differential_eq	Dynamical	SSE_6	no

In Table 5.3 the best model is M_4 because:

$$SSE_4 = \min \{SSE_1, SSE_2, SSE_3, SSE_4, SSE_5, SSE_6\}$$

The implementation of this minimization procedure is easy once the numerical values of the Sum of the Squares (SSE) are calculated by the numerical module.

We have to say here that this method for selecting the "best" model for a given problem can be improved in several ways to consider other factors that relate to this decision process. For example, one may like to consider the "type" of the model or the "simplicity" of the model as other factors of importance in the process of "best" model selection. In this case, a set of if-then rules would be required to make the decision and this module would be then considered a real "knowledge base". For the moment, we have only a method for "best" model selection that uses statistical measures and "knowledge" about the process of mathematical modelling.

5.4 Comparison with Related Work

There has been some work recently in the area of numerical law discovery, but much of the research in Machine Learning is in other areas such as induction (Sleeman & Edwards, 1992). We think that this is mainly because "discovery" is a more difficult kind of "learning" (in this case, finding the best mathematical models for a given data set). However, we can say that automated mathematical modelling is very important for many domains of application for obvious reasons. For example, in the engineering and robotics domains is critical to obtain mathematical models for the problems, to be able to understand them and also to be able to predict and control their future behavior.

Similar work with respect to Machine Learning can be found in a paper by Moulet (1992), however the approach to model discovery is different that the one presented here (this can be seen from the heuristic method proposed by Moulet). Also in a paper by Rao and Lu (1993) we can see a method for model discovery for engineering domains, but also with a different approach (his approach is similar to "clustering"). Also, there is another very important difference with other authors, in the kind of mathematical models that we are considering for our intelligent system. We are considering non-linear mathematical models from the theory of Dynamical Systems and not only linear regression models like other authors. This is because non-linear dynamical models offer the possibility of explaining the erratic behavior of real time series with "chaos theory" (Devaney, 1989).

5.5 Summary

We have presented in this chapter a new method for automated mathematical modelling of non-linear dynamical systems. This method is based on a hybrid fuzzy-fractal approach to achieve, in an efficient way, automated modelling for a particular problem using a time series as a data set. The use of the fractal dimension is to perform time series analysis of the data, so as to obtain a qualitative characterization of the time series. The use of fuzzy logic techniques is to simulate the process of expert model selection using the qualitative information obtained from the time series analysis module. At the end, the "best" mathematical model is obtained by comparing the measures of goodness for the selected mathematical models. In Chapter 8, we will explore some advanced applications of this method for automated mathematical modelling. The results will show the efficiency and potential benefits of using this new method for modelling and simulation of complex non-linear dynamical systems.

Chapter 6

Fuzzy-Genetic Approach for Automated Simulation

This Chapter describes the important problem of numerical simulation for non-linear dynamical systems and its solution by using intelligent methodologies. The numerical simulation of a particular dynamical system consists in the successive application of a map (difference equation) and the subsequent identification of the corresponding dynamic behaviors. Automated simulation of a given dynamical system consists in selecting the appropriate parameter values for the mathematical model and then applying the corresponding iterative method (map) to find the limiting behavior. In this chapter, a new method for automated parameter selection based on the use of genetic algorithms, is introduced. Also, a new method for dynamic behavior identification based on fuzzy logic, is introduced. The fuzzy-genetic approach for automated simulation consists in the integration of the method for automated parameter selection (based on GA) and the method for behavior identification (based on fuzzy logic).

6.1 The Problem of Automated Simulation

In this section, we describe briefly the problem of numerical simulation for non-linear dynamical systems. First, we present some basic concepts and the main goal of the field of numerical simulation. Then, we present some basic concepts about dynamical systems theory that we consider necessary to understand the methods that will be described later in the chapter. At the end, we describe the problem of automated simulation of dynamical systems

6.1.1 Numerical simulation of dynamical systems

Real dynamical systems can be represented by mathematical models expressed as non-linear differential equations of the form:

$$dY/dt = f(Y, t, \theta) \tag{6.1}$$

where Y is a vector of dynamic variables, t is time, and θ is a vector of parameters, or as non-linear difference equations of the form:

$$Y_{t+1} = F(Y_t, Y_{t-1}, ..., \theta) \tag{6.2}$$

In the case of Equation (6.1) one says that the model is a continuous one. On the other hand, for Equation (6.2) the model is considered of the discrete type.

The simulation of the real dynamical system, in the first case, consists in the numerical solution of the non-linear differential equation along with the identification of all the possible corresponding dynamic behaviors of the system. For example, the numerical solution of the differential equation can be obtained by the well known second-order Runge-Kutta method (Nakamura, 1997):

$$Y_{n+1} = Y_n + 1/2(k_1 + k_2)$$
$$k_1 = h f(Y_n, t_n, \theta) \tag{6.3}$$
$$k_2 = h f(Y_n, k_1, t_{n+1}, \theta)$$

where h is the stepsize of the method. Equation (6.3) can be used for different parameter values of θ to obtain the different dynamic behaviors of the system described by the model of Equation (6.1). Of course, more advanced methods for the numerical solution of differential equations can be used if more accuracy is desired.

The simulation of the real dynamical system, in the second case, consists in the successive application of the map given by Equation (6.2) for different parameter values of θ and then identifying the different corresponding dynamic behaviors of the system.

In any case, the result of the numerical simulation of the real dynamical system (represented by Eq. (6.1) or Eq. (6.2)) is a time series of the following form:

$$Y_1, Y_2, Y_3, ..., Y_p \quad \text{for } \theta = \theta* \tag{6.4}$$

this time series represents the motion of the dynamical system for the specific parameter value $\theta = \theta*$. It is well known in dynamical systems theory, that this time series can have many different types of dynamic behaviors ranging from very simple periodic fixed points to the very complicated "chaotic" behavior. Even more, for non-linear mathematical models it is possible to have all the range of dynamic behaviors for different parameter values of θ. For this reason, the numerical simulation of relative complex non-linear mathematical models requires a lot of exploration in order to find all of the possible dynamic behaviors for the real dynamical system.

6.1.2 Behavior identification for dynamical systems

There are theoretical results that can be used to establish the existence of chaotic trajectories or other dynamic behaviors for several dynamical systems (Devaney, 1989). However, in many cases it may be difficult or analytically impossible to detect a period-three cycle (required by Li/Yorke's Theorem) necessary to identify chaotic behavior, and for most differential equation systems there are no theoretical results at all. Experiments show that even for cycles of a relatively low period it may be impossible to distinguish regular time series from completely chaotic time series by simple visual inspection.

If a dynamical system is given whose behavior can not be investigated further by applying the standard geometric or analytical methods, numerical simulations are appropriate. The generated time series in such a simulation may exhibit simple patterns like monotonic convergencies or harmonic oscillations. However, the series may also appear to be random due either to

- periodic behavior with a long period
- quasiperiodic behavior with many different frequencies

- deterministic chaos, or to
- noise generated by the use of specific algorithms during the simulation.

The following numerical tools can be useful in deciding whether an actual time series generated by the simulation of a known dynamical system is regular, chaotic, or stochastic.

Spectral analysis: Has proven to be particularly useful in attempts to distinguish periodic and quasi-periodic time series with few frequencies from random behavior. The aim of spectral analysis is dividing a given time series into different harmonic series with different frequencies. For example, if a time series consists of two overlapping harmonic series, spectral analysis attempts to isolate these two harmonic series and to calculate the involved frequencies. Furthermore, spectral analysis provides information on the contribution of each harmonic series to the overall motion. While power spectra are particularly useful in investigating the periodic or quasi-periodic behavior of dynamical systems, chaotic and random behavior can not be discriminated with this method. For this reason, we will not describe this statistical method in more detail.

The short presentation of spectral analysis has shown that traditional statistical techniques fail to provide a definite answer to the question of whether a given complex time series is generated by a random process or by deterministic laws of motion. Appropriate concepts for distinguishing between these two sources of complex and irregular behavior have emerged only recently, and the development of new techniques is still in progress. In addition to the empirical motivation for dealing with these concepts, their discussion will be useful because new insights into the nature of deterministic chaotic systems can be provided.

Phase space embedding: Of central importance to the numerical investigation of complex dynamical systems is the notion of the "embedding dimension". Suppose that a dynamical process is generated by a deterministic set of equations like

$$x^i_{t+1} = g_i(x_t), \qquad x \in R^n, \qquad i = 1, ..., n \qquad (6.5)$$

and let a certain x^j be the variable which attracts the attention of an observer. The observer neither knows the structural form of (6.5) and its dimension n, or can he be sure that this measurement of the quantity x^j_t is correct. Denote the observed value of the variable x^j at t as $\bar{x}^j_t$ and let

$$\bar{x}^j_t = h(x_t) \qquad\qquad (6.6)$$

i.e., the observed variable depends on the "true" values x^i_t, but the measurement of the variable may imply differences between $\bar{x}^j_t$ and x^j_t.

The measurement procedure over time generates a time series $\{ \bar{x}^j_t \}^T_{t=1}$. An "embedding" is an artificial dynamical system which is constructed from the one-dimensional time series in the following way: consider the last element $\bar{x}^j_T$ in the observed time series and combine it with its m predecessors into a vector $\bar{x}^m_T$ $= (\bar{x}^j_T , \bar{x}^j_{T-1} , \bar{x}^j_{T-m+1})$. Perform this grouping for every element $\bar{x}^j_t$ in the descending order t = T, ..., 1 and drop the remaining m-1 first elements in the original time series because they do not possess measured predecessors. The m-dimensional vector $\bar{x}^m_T$ is called the "m-history" of the observation $\bar{x}^j_t$. Since the first elements are dropped, the sequence of the vectors $\{ \bar{x}^m_t \}^T_{t=t_0}$ is shorter than the original time series and varies with the length of the history. The length m is called the "embedding dimension".

Each m-history describes a point in an m-dimensional space, the coordinates of which are the delayed observed values in the vector $\bar{x}^m_t$. The sequence $\{ \bar{x}^m_t \}^T_{t=t_0}$ of points will therefore form a geometric object in this space. It was proven by Takens (1981) that this object is topologically equivalent to the appropriate object generated by the true dynamical system (6.5) if the functions g_i and h are smooth and m > 2n-1. If these conditions are satisfied, it is thus theoretically possible to reconstruct the behavior of the (unknown) true dynamical system from a single observed time series.

Correlation dimension: Suppose that an attractor is chaotic and consider two points on this attractor which are far apart in time. Due to the sensitive dependence on initial conditions, these points are dynamically uncorrelated since arbitrarily small measurement errors in the determination of the initial point can lead to drastically different locations of the second point. However, as both points are located on an attractor, they may come close together in phase space, i.e., they may be spatially correlated.

The two points $\bar{x}^{mi}$, $\bar{x}^{mj}$ are said to be spatially correlated if the Euclidian distance is less than a given radius r of an m-dimensional ball centered at one of the two points, i.e., $\| \bar{x}^{mi} - \bar{x}^{mj} \| < r$. The spatial correlation between all points on the attractor for a given r is measured by

$$c(r,m) = \lim_{T \to \infty} \frac{1}{T^2} \times [\text{number of pairs i,j with distance } \| \bar{x}^{mi} - \bar{x}^{mj} \| < r] \quad (6.7)$$

The function c(r,m) is called the "correlation integral". The "correlation dimension" is defined as:

$$D^c(m) = \lim_{r \to 0} \frac{\ln c(r,m)}{\ln r} \quad (6.8)$$

The calculated values of the correlation dimension are close to the Hausdorff-Besicovitch dimension and do not exceed it. Obviously, the correlation dimension can be computed more easily than the Hausdorff-Besicovitch dimension since counting is the essential ingredient in calculating the correlation dimension: fix a small r and count the number of points N(r) lying in a ball centered at a $\bar{x}^{mi}$. Perform this procedure for every $\bar{x}^{mi}$ and calculate c(r,m) and $D^c(m)$.

Lyapunov exponents: Strange attractors are geometrically characterized by the simultaneous presence of "stretching" and "folding", implying that two initially close points will be projected to different locations in phase space. The presence and interaction of stretching and folding in a certain dynamical system can be described via the so-called "Lyapunov exponents". The Lyapunov exponents constitute a quantity for characterizing the rate of divergence of two initial points (Rasband, 1990). Note that this divergence on the attractor is a dynamical property. Consider first the discrete-time case with an n--dimensional mapping

$$x_{t+1} = f(x_t), \qquad x \in R^n \quad (6.9)$$

and two initial points xo and x'o. Let the difference $\delta x_0 = x_0 - x'_0$ be small. After, the first iteration, the difference between the two points will be

$$x_1 - x'_1 = f^{(1)}(x_0) - f^{(1)}(x'_0) \quad (6.10)$$

A linear approximation of the difference yields

$$x_1 - x'_1 \approx \frac{d\ f^{(1)}(x_0)}{dx} \delta x_0 \tag{6.11}$$

where $d\ f^{(1)}(x_0)\ /dx$ is the Jacobian matrix J. After N iterations the difference between the corresponding points will be

$$x_N - x'_N = f^{(N)}(x_0) - f^{(N)}(x'_0) \tag{6.12}$$

and linearization yields

$$x_N - x'_N \approx \frac{d\ f^{(N)}(x_0)}{dx} \delta x_0 \tag{6.13}$$

where, by the chain rule, $df^{(N)}(x_0)\ /\ dx = J^{(N)}$ equals the product of N Jacobian matrices J evaluated at x_0.

As $J^{(N)}$ is an nxn matrix, it also possesses n eigenvalues. Denote the eigenvalues of this matrix as Λ^N_i and rearrange them such that $\Lambda^N_1 \geq \Lambda^N_2 \geq ...\Lambda^N_n$. The Lyapunov exponents $\lambda i;\ i = 1, ...,n$, are defined as

$$\lambda i = \lim_{N \to \infty} \frac{1}{N} \log_2(\Lambda^N_i) \tag{6.14}$$

An analogous procedure for the continuous-time case leads to

$$\lambda i = \lim_{T \to \infty} \frac{1}{T} \log_2(\Lambda^T_i) \tag{6.15}$$

with $T \in R$, i.e., the time step between iterations tends to zero.

The meaning of the Lyapunov exponents can be interpreted as follows: when all Lyapunov exponents are negative on an attractor, the attractor is an asymptotically stable fixed point. When one or more Lyapunov exponents are non-negative, then at least one exponent must vanish. A limit cycle must involve a $\lambda i = 0$ and thus cannot occur in the one-dimensional case. A torus can emerge only in at least three-dimensional phase space. As two cyclical directions are involved in a 2-torus, two of its Lyapunov exponents are equal to zero. If one of the exponents is positive, chaotic motion prevails (Rasband, 1990).

The characterization of the behavior of low-dimensional continuous-time dynamical systems by means of their Lyapunov exponents is summarized in Table 6.1. Empty fields indicate the impossibility of the appropriate dynamical behavior if the dimension n is two low.

Table 6.1 Lyapunov exponents and Dynamical Behavior in
          Continuous-Time Systems

Dimension	Asymptotic Stability	Limit cycle (T^1)	Torus (T^2)	Chaos
n = 1	(-)			
n = 2	(-, -)	(0, -)		
n = 3	(-, -, -)	(0, -, -)	(0, 0, -)	(+, 0, -)

6.1.3 Automated simulation of dynamical systems

The problem of performing an efficient simulation for a particular dynamical system can be better understood if we consider a specific mathematical model. Let us consider the following model:

$$X' = \sigma(Y-X)$$
$$Y' = rX - Y - XZ \qquad\qquad (6.16)$$
$$Z' = XY - bZ$$

where X, Y, Z, σ, r, b $\in$ R, and σ, r and b are three parameters which are normally taken, because of their physical origins, to be positive. The equations are often studied for different values of r in $0 < r < \infty$. This mathematical model has been studied by Rasband (1990) to some extent, however there are still many questions to be answer for this model with respect to its very complicated dynamics for some ranges of parameter values.

If we consider simulating Equation (6.16), for example, the problem is of selecting the appropriate parameter values for σ, r, b, so that the interesting dynamical behavior of the model can be extracted. The problem is not an easy one, since we need to consider a three-dimensional search space σ r b and there are many possible dynamical behaviors for this model. In this case, the model consisting of three simultaneous differential equations, the behaviors can range from simple periodic orbits to very complicated chaotic attractors. Once the parameter values are selected then the problem becomes a numerical one, since then we need to iterate an appropriate map to approximate the solutions numerically.

The problem of performing automated simulation for a particular dynamical system is then of finding the "best" set of parameter values for the mathematical model. Our general algorithm (Castillo & Melin, 1995c) for selecting the "best" set of parameter values is shown in Figure 6.1.

The algorithm shown in Figure 6.1 can be explained as follows: first, the mathematical model is analyzed to "understand" it; second, a set of admissible parameters is generated using the understanding of the model; third, a specific genetic algorithm is used to select the best set of parameter values; finally, the numerical simulations are performed and the dynamical behaviors are identified using fuzzy logic.

STEP 1 Read the mathematical model M.

STEP 2 Analyze the model M to "understand" its complexity.

STEP 3 Generate a set of admissible parameters using the understanding of the model.

STEP 4 Perform a selection of the "best" set of parameter values. This set is generated using a specific genetic algorithm.

STEP 5 Perform the simulations by solving numerically the equations of the mathematical model. At this time the different types of dynamical behaviors are identified using a fuzzy rule base.

Figure 6.1 New algorithm for selecting the best set of parameter values.

The implementation of the new method for Automated Simulation as a computer program was done using the PROLOG programming language (the complete program is shown in Appendix B). The choice of PROLOG is because of its symbolic manipulation features and also because it is an excellent language for developing prototypes (Bratko, 1990). The general architecture of the prototype intelligent system for simulation is the same as the one shown in Figure 5.3. The main difference is that the knowledge base for simulation now consists of two modules: Parameter Selection, and Dynamic Behavior Identification. In the following lines we will describe these two modules in more detail.

6.2 Method for Automated Parameter Selection using Genetic Algorithms

The knowledge for simulation of the intelligent system consists in the application of a specific genetic algorithm (Jang, Sun & Mizutani, 1997) to select the best set of parameters of a particular dynamical system. Our genetic algorithm for parameter value selection (Castillo & Melin, 1998b) can be defined as shown in Figure 6.2.

STEP 1 Initialize a population with randomly generated individuals
        (parameters) and evaluate the fitness value of each individual

STEP 2 (a) Select two members from the population with probabilities
        proportional to their fitness values

      (b) Apply crossover with a probability equal to the crossover rate

      (c) Apply mutation with a probability equal to the mutation rate

      (d) Repeat (a) to (d) until enough members are generated to form the
        next generation

STEP 3 Repeat steps 2 and 3 until the stopping criterion is met

---

Figure 6.2 Genetic algorithm for parameter value selection.

The fitness function should evaluate the dynamical information given by a particular set of parameter values, i.e. the fitness function should measure the power of the parameter set. Lets consider a three-dimensional model with 3 parameters θ, α and γ, then assuming that we have only four possible dynamical behaviors (for a given system):

> B0: fixed point of period 1
> B1: fixed point of period 2
> B2: fixed point of period 4
> B3: fixed point of period 8
> B4: chaotic behavior

we will have that the parameter set $\pi = (\theta, \alpha, \gamma)$, where $\theta, \alpha, \gamma \in R$, can result in any of the five possible behaviors. In this case, we need to consider 5 individuals in the Population and an initial population can be denoted as:

$$P_0 = (\pi_{01}, \pi_{02}, \pi_{03} \; \pi_{04} \; \pi_{05})$$

For an initial population there is a high probability that most of the π_i could give the B0 behavior, so there has to be evolution to obtain a better parameter set. The identification of the respective behaviors can be done by iteration of the dynamic systems or by other mathematical means, for example the fractal dimension or the Lyapunov exponents (Rasband, 1990). The fitness value of each individual in population P_i can by defined as follows (Castillo & Melin, 1998a):

$$
\begin{aligned}
F(\pi_{ij}) &= 1 \quad \text{for fixed point of period 1} \\
F(\pi_{ij}) &= 2 \quad \text{for fixed point of period 2} \\
F(\pi_{ij}) &= 4 \quad \text{for fixed point of period 4} \\
F(\pi_{ij}) &= 8 \quad \text{for fixed point of period 8} \\
F(\pi_{ij}) &= 10 \quad \text{for chaotic behavior}
\end{aligned}
$$

this is only one of the possible schemes that can be used for this case. In this case, we have assigned the fitness values proportional to the complexity of the dynamic behavior to guide the search of the genetic algorithm. However, the specific numeric values could be changed to suit the needs of particular applications. We need to remember here that the fitness function has to be designed for each specific application of a genetic algorithm.

A more general form for defining the fitness function for real dynamical systems can be establish by using the fractal dimension d_f of the time series generated by the numerical simulation of the dynamical system. Mathematically, we can define the fitness function as:

$$F(\pi_{ij}) = e^{\ df(\pi ij)}, \qquad 0 \le d_f \le 3, \tag{6.17}$$

where $d_f(\pi_{ij})$ = fractal dimension of the time series for parameter set π_{ij}. The general idea of Equation (6.17) is to assign a bigger value to the fitness function when the complexity of the time series, generated by the simulation, is greater (which is true, of course, when d_f is of a higher value). Of course, here the use of the exponential function is only to spread the values of the fractal dimension but other functions could be used as well.

6.3 Method for Dynamic Behavior Identification using Fuzzy Logic

Once the parameter values have been found and the numerical simulations have been performed then the final step is to identify the possible dynamic behaviors of the system. The knowledge for behavior identification can be expressed as a fuzzy-rule base that uses the information obtained in the numerical simulation to identify the different behaviors of the model. To give an idea of how this knowledge can be expressed as a fuzzy-rule base we show below two sample schemes that can be used for behavior identification.

6.3.1 Behavior identification based on the analytical properties of the model

We can build a set of fuzzy rules for dynamic behavior identification based on the analytical properties of the mathematical models and using the well known theorems of dynamical systems theory (Castillo & Melin, 1997b). To give an idea of how this knowledge can be translated to fuzzy rules we show below some sample rules for several types of dynamical systems.

1) <u>Single-link Robot Model</u>: This mathematical model of a sinusoidally non-linear robot consist of two simultaneous differential equations:

$$q' = Q \hspace{4cm} (6.18)$$
$$Q' = (K_t I - N\sin(q) - F_d Q) / M_q$$

where the parameters I, M_q, N, F_d and K_t are all positive. Lim, Hu and Dawson (1996) have presented an extensive gallery of periodic and a-periodic motions for this model. In this case the equilibria (q^*, Q^*) is stable if and only if the real parts of the eigenvalues are negative and this is equivalent to the rule:

IF   a > 0   **THEN**   Equilibria = stable

where a is defined by the characteristic equation:

$$\lambda^2 + a\lambda + b = 0$$

with a = - trJ, b = detJ. Where "trJ" is the trace and "detJ" is the determinant of the Jacobian Matrix.

2) <u>Other Bi-dimensional Models</u>: Similar bi-dimensional autonomous models can be written in the following manner:

$$X' = \alpha \, f(X,Y) \hspace{4cm} (6.19)$$
$$Y' = \beta \, g(X,Y)$$

In this case, the Equilibria (X^*, Y^*) is stable if:

$$\alpha \, f_x + (g_y - \beta) < 0$$

where f_x and g_y are partial derivatives. In fuzzy logic language we have the following rule:

IF $[\alpha \, f_x + (g_y - \beta) < 0]$ **THEN**   Equilibria = stable

Also we have the following rule for a Hopf Bifurcation:

IF $\alpha_0 = (\beta - g_y)/f_x$ **THEN**   Hopf_Bifurcation

which gives us the condition for a Hopf bifurcation to occur.

3) <u>Firth's Model of a single-mode laser</u>: The basic equations of a single-mode (unidirectional) homogeneously broadened laser in a high-finesse cavity, tuned to resonance, may be written as a system of three differential equations (Abraham & Firth, 1984):

$$X' = \gamma_c \, (X + 2C_p)$$
$$P' = - \, \Gamma \, (P - XD) \tag{6.20}$$
$$D' = - \, \gamma \, (D + XP - 1)$$

Here X is a scaled electric field (or Rabi frequency), γ_c is a constant describing the decay of the cavity field and C is the cooperativity parameter.

In this case, the Equilibria (X^*,P^*,D^*) is stable if a, b, c > 0 and (ab - c) > 0, where a, b and c are defined by the characteristic equation for the system. We can also have more complicated rules for other types of dynamical behaviors.

4) <u>Other three-dimensional Models</u>: A three-dimensional system of differential equations can be written in the following form:

$$X' = \alpha f(X,Y,Z)$$
$$Y' = \beta g(X,Y,Z) \tag{6.21}$$
$$Z' = \gamma h(X,Y,Z)$$

In this case, the Equilibria (X^*,Y^*,Z^*) is stable if a, b, c > 0 and (ab - c) > 0, where a, b and c are defined by the characteristic equation for the system:

$$\lambda^3 + a \, \lambda^2 + b \, \lambda + c = 0$$

In fuzzy logic language we have the rule:

IF	a,b,c >0	**AND**	(ab-c) >0
		THEN	Equilibria = stable

other rules follow in the same manner for all the types of dynamical behaviors possible for this class of mathematical models.

We have to note here that in this case the computer program for this method needs to obtain the symbolic derivatives for the functions in the conditions of the rules. This is critical for the problem of behavior identification, since we require these derivatives to obtain the values of the parameters in the rules. This will make this method time consuming because the time series from the simulations are not used at all.

6.3.2 Behavior identification based on the fractal dimension and the Lyapunov exponents

We can obtain a more efficient method of dynamic behavior identification, if we make use of the information contained in the time series that resulted from the simulation of the dynamical system. From the time series of the numerical simulations, we can calculate the Lyapunov exponents of the dynamical system and also the fractal dimension of the time series. With this dynamical information, we can easily identify the corresponding behaviors of the system.

For dissipative dynamical systems, for example, we can use the results that where shown in Table 6.1 to build a set of fuzzy rules for behavior identification using the Lyapunov exponents. However, since the Lyapunov exponents can only identify between asymptotic stability, general limit cycles and chaos, we need to use the fractal dimension d_f to discriminate between the different periodic behaviors possible. Based on prior empirical work (Castillo & Melin, 1996b), we have been able to use the fractal dimension to discriminate between different periodic behaviors. Then, if we combine the use of the Lyapunov exponents with the use of the fractal dimension, we can obtain a set of fuzzy rules that can identify in a one-to-one manner the different dynamic behaviors. The if-then rules have to be "fuzzy" because there is uncertainty associated with the numerical values of the Lyapunov exponents and also the classification scheme (for the limit cycles) using the fractal dimension is only approximated.

We show in Table 6.2 the fuzzy rule base that we have developed for dynamic behavior identification for dynamical systems of up to three variable. The empty fields in Table 6.2 indicate no use of the fractal dimension for that case.

We can define membership functions for the numerical intervals of the fractal dimension, for the Lyapunov exponents and for the behavior identifications shown in Table 6.2. Once this membership functions are defined, the usual fuzzy reasoning methodology can be applied to implement this method of behavior identification.

Table 6.2 Fuzzy rule base for behavior identification using Lyapunov exponents and fractal dimension

IF			THEN
Number of variables	Lyapunov exponents	Fractal Dimension	Behavior Identification
1	(-)		stable fixed point
2	(-, -)		stable fixed point
2	(0, -)	[1.1, 1.2)	limit cycle of period 2
2	(0, -)	[1.2, 1.3)	limit cycle of period 4
2	(0, -)	[1.3, 1.4)	limit cycle of period 8
2	(0, -)	[1.4, 1.5)	limit cycle of period 16
3	(-, -, -)		stable fixed point
3	(0, -, -)	[2.1, 2.2)	limit cycle of period 2
3	(0, -, -)	[2.2, 2.4)	limit cycle of period 4
3	(0, -, -)	[2.4, 2.6)	limit cycle of period 8
3	(0, -, -)	[2.6, 2.8)	limit cycle of period 16
3	(+, 0, -)	[2.8, 3.0)	chaos

6.4 Summary

We have presented in this chapter a new method for automated simulation of non-linear dynamical systems. This method is based on a hybrid fuzzy-genetic approach to achieve, in an efficient way, automated simulation for a particular dynamical system given its mathematical model. The use of genetic algorithms is to achieve automated parameter selection for the models. The use of fuzzy logic is to simulate the process of expert behavior identification by implementing the knowledge of identification by a set of fuzzy rules. In Chapter 8, we will explore some advanced applications of this method for automated simulation. The results will show the efficiency of this new method for the simulation of complex non-linear dynamical systems.

Chapter 7

Neuro-Fuzzy Approach for Adaptive Model-Based Control

We describe in this chapter a new method for adaptive control of Non-Linear Dynamical Systems based on the use of Neural Networks, Fuzzy Logic and mathematical models. Dynamical Systems can have many forms and be of many types, but one very important case is that of non-linear dynamic plants. Production processes in real world Plants are often highly non-linear and difficult to control. The problem of controlling them using conventional controllers has been widely studied (Albertos, Strietzel & Mart, 1997). Much of the complexity in controlling any process comes from the complexity of the process being controlled. This complexity can be described in several ways. Highly non-linear systems are difficult to control, particularly when they have complex dynamics (such as instabilities to limit cycles and chaos). Difficulties can often be presented by constraints, either on the control parameters or in the operating regime. Lack of exact knowledge of the process, of course, makes control more difficult. Optimal control of many processes also requires systems which make use of predictions of future behavior. The mathematical models for the Plants are assumed to be expressed as systems of differential equations. The goal of having these models is to capture the dynamics of production processes, so as to have a way of controlling this dynamics for industrial purposes. Accordingly, this chapter is divided into four parts: Modelling the Process of the Plant, Neural Networks for Control, Fuzzy Logic for Model Selection, and Neuro-Fuzzy Adaptive Model-Based Control. Of course, even if we illustrate here our new method for adaptive control only for non-linear plants, the method can also be used for general non-linear dynamical systems.

7.1 Modelling the Process of the Plant

The problem of automated mathematical modelling for non-linear dynamical systems was considered in Chapter 5 of this book. However, we need to consider the problem of modelling related to achieving model-based control. In this case, we need mathematical models of the specific dynamical system to have a reference dynamic behavior that the controller can follow. Also, it is important to consider the control parameters in the mathematical model, so that we can apply the appropriate control actions to the dynamical system. We will illustrate in this section these ideas for the case of dynamic plants and will show in the following sections how to use the models for adaptive control.

We need a mathematical model of the non-linear dynamic plant to understand the dynamics of the processes involved in production. For a specific case, this may require testing several models before obtaining the appropriate mathematical model for the process. For real world plants with complex dynamics, we may even need several models for different set of parameter values to represent all of the possible behaviors of the plant. Mathematical models for the plants can be expressed as differential equations (in continuous time) or alternatively as difference equations (in discrete time). We assume in this chapter (without loss of generality) that the models are expressed as differential equations.

The simplest mathematical model for a non-linear plant can be expressed as follows:

$$dx/dt = f_1(x) - \beta f_2(x) \tag{7.1}$$
$$dp/dt = \beta f_2(x)$$

where x = state variable, p = quantity of the product, β = constant dependent on the efficiency of the conversion process (production). This is a good mathematical model for the case in which there is only one input into the production process (x) and there is only one product (p). Functions $f_1(x)$ and $f_2(x)$ should represent the dynamics of the plant.

For the case of two inputs to the process x_1, x_2 and one product p, we have the following general mathematical model:

$$dx_1/dt = f_1(x_1, x_2) - \beta f_2(x_1) \qquad\qquad (7.2)$$
$$dx_2/dt = g_1(x_1, x_2) - \gamma g_2(x_2)$$
$$dp/dt = \beta f_2(x_1) + \gamma g_2(x_2)$$

where β and γ are constants measuring the efficiency of the production process. In this case, we have to design four functions $f_1(x_1, x_2)$, $f_2(x_1)$, $g_1(x_1, x_2)$ and $g_2(x_2)$ to represent the corresponding non-linear plant.

For the case of two inputs to the process x_1, x_2, one desired product p, and one undesirable product x_3, we can have the following mathematical model:

$$dx_1/dt = f_1(x_1, x_2) - \beta f_2(x_1) - \sigma_1 x_1 x_3 \qquad\qquad (7.3)$$
$$dx_2/dt = g_1(x_1, x_2) - \gamma g_2(x_2) - \sigma_2 x_2 x_3$$
$$dx_3/dt = h(x_3) + \sigma_1 x_1 x_3 + \sigma_2 x_2 x_3$$
$$dp/dt = \beta f_2(x_1) + \gamma g_2(x_2)$$

where β, γ, σ_1 and σ_2 are constants measuring the efficiency of the production process. In this case, we have to design five functions $f_1(x_1, x_2)$, $f_2(x_1)$, $g_1(x_1, x_2)$, $g_2(x_2)$ and $h(x_3)$ to represent the corresponding plant.

For the case of two inputs to the process x_1, x_2 and two products p_1 and p_2, we have the general mathematical model:

$$dx_1/dt = f_1(x_1, x_2) - \beta_1 f_2(x_1) - \beta_2 f_3(x_1) \qquad\qquad (7.4)$$
$$dx_2/dt = g_1(x_1, x_2) - \gamma_1 g_2(x_2) - \gamma_2 g_3(x_2)$$
$$dp_1/dt = \beta_1 f_2(x_1) + \gamma_1 g_2(x_2)$$
$$dp_2/dt = \beta_2 f_3(x_1) + \gamma_2 g_3(x_2)$$

where β_1, β_2, γ_1, γ_2 are constants measuring the efficiency of the production process. In this case, we have to design six functions $f_1(x_1, x_2)$, $f_2(x_1)$, $f_3(x_1)$, $g_1(x_1, x_2)$, $g_2(x_2)$ and $g_3(x_2)$ to represent the corresponding plant.

General mathematical models for more complicated non-linear plants can be developed in a similar manner. In any case, the models can always be represented as sets of coupled simultaneous non-linear differential equations.

The control of non-linear plants, represented by mathematical models like the ones described before, is a very complicated task because of the complex dynamics that can arise. The mathematical models, given by Equations (7.2), (7.3) and (7.4), can exhibit a wide range of dynamical behaviors (from periodic ones to even chaotic behavior). For this reason, it is very important to design control methods that can learn to control (in an adaptive manner) complex dynamical systems. These methods should use the knowledge and information about the plant contained in the mathematical models and also should use "intelligent" methodologies to really achieve adaptive and learning capabilities. In the following sections of this chapter we will describe in more detail how can we achieve this model-based adaptive control of non-linear plants.

7.2 Neural Networks for Control

Parametric Adaptive Control is the problem of controlling the output of a dynamical system with a known structure but unknown parameters. These parameters can be considered as the elements of a vector **p**. If p is known, the parameter vector θ of a controller can be chosen as θ^* so that the plant together with the fixed controller behaves like a reference model described by a differential equation with constant coefficients (Narendra & Annaswamy, 1989). If p is unknown, the vector $\theta(t)$ has to be adjusted on-line using all the available information concerning the dynamical system.

Two distinct approaches to the adaptive control of an unknown plant are (i) direct control and (ii) indirect control. In direct control, the parameters of the controller are directly adjusted to reduce some norm of the output error. In indirect control, the parameters of the plant are estimated as $\hat{p}(t)$ at any time instant and the parameter vector $\theta(t)$ of the controller is chosen assuming that $\hat{p}(t)$ represents the true value of the plant parameter vector. Even when the plant is assumed to be linear and time-invariant, both direct and indirect adaptive control result in non-linear systems.

When the plant is non-linear and dynamic (i.e. the present value of its output depends upon the past values of the input and the output respectively), a neural network can be used as a controller as shown in Figure 7.1. This corresponds to direct control.

Direct Control: In conventional direct adaptive control theory, methods for adjusting the parameters of a controller based on the measured output error rely on concepts such as positive realness and/or passivity. By making suitable assumptions concerning the plant and the reference model, it is shown that the direction in which a parameter is to be adjusted can be obtained by correlating two signals that can be measured.

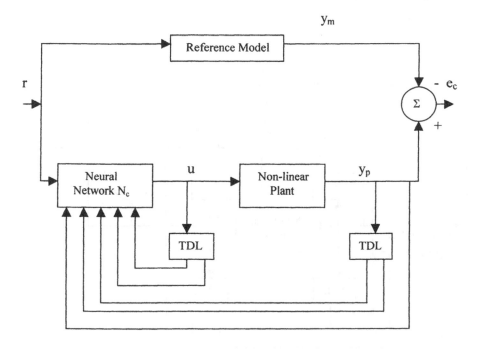

Figure 7.1 Direct adaptive control using neural networks.

At present, methods for directly adjusting the parameters of the controller (the neural network N_c in Figure 7.1) in a stable fashion based on the output error are not available. This is due to the non-linear nature of both the plant and the controller. Even backpropagation cannot be used directly, since the plant is unknown and hence cannot be used to generate the desired partial derivatives. Hence, until direct control methods are developed, adaptive control of non-linear dynamical systems has to be carried out using indirect control methods.

Indirect Control: As mentioned earlier, when indirect control is used to control a non-linear system, the plant is parameterized using one of the models described in the previous section and the parameters of the model are updated using the identification error. The controller parameters in turn are adjusted by backpropagating the error (between the identified model and the reference model outputs) through the identified model. A block diagram of such an adaptive system is shown in Figure 7.2.

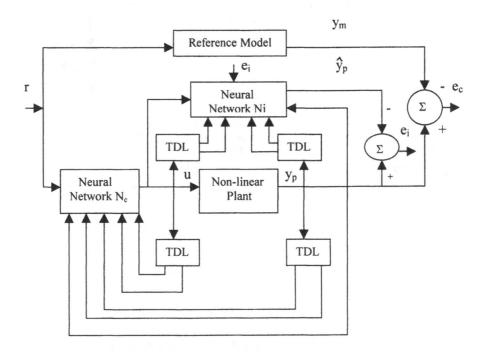

Figure 7.2 Indirect adaptive control using neural networks.

Both identification and control can be carried out at every instant or after processing the data over finite intervals. When external disturbances and/or noise are not present in the system, it is reasonable to adjust the control and identification parameters synchronously. However, when sensor noise or external disturbances are present, identification is carried out at every instant while control parameter updating is carried out over a slower time scale, to assure robustness.

The structure of the adaptive system proposed in this work to control a non-linear dynamic plant is similar to the one shown in Figure 7.2, the main difference is that we use a decision scheme to select the appropriate reference model for the plant. This decision scheme is based on the use of fuzzy logic techniques and is explained in the next section. In Figure 7.2, the delayed values of the plant input and plant output form the inputs to the neural network N_c which generates the feedback control signal to the plant. The parameters of the neural network N_i are adjusted by backpropagating the identification error e_i while those of the neural network N_c are adjusted by backpropagating the control error (between the output of the reference model and the identification model) through the identification model.

Process control of non-linear plants is an attractive application because of the potential benefits to both adaptive network research and to actual process control in the industry. In spite of the extensive work on self-tuning controllers and model-reference control, there are many problems in the real world processing industries for which current techniques are inadequate. Many of the limitations of current adaptive controllers arise in trying to control poorly modeled non-linear systems. For most of these processes extensive data are available from past runs, but it is difficult to formulate precise models. This is precisely where adaptive networks are expected to be useful (Ungar, 1995).

7.3 Fuzzy Logic for Model Selection

For a complex dynamical system it may be necessary to consider a set of mathematical models to represent adequately all of the possible dynamic behaviors of the system (Melin & Castillo, 1997). In this case, we need a decision scheme to select the appropriate model to use according to the value of a selection parameter α. In this section, we show a method for model selection based on fuzzy logic and a new fuzzy inference system for differential equations.

We have designed a method, based on fuzzy logic techniques, for mathematical model selection using as input the numerical value of a selection parameter α. We assume, in what follows, that parameter α is defined over a real-valued interval:

$$\alpha_0 \leq \alpha \leq \alpha_n . \tag{7.5}$$

We also assume that we have n mathematical models considered appropriate for the respective n subintervals, defined on $[\alpha_0, \alpha_n]$, as follows:

$$\alpha_0 \leq \alpha < \alpha_1 , \qquad \alpha_1 \leq \alpha < \alpha_2 , ..., \qquad \alpha_{n-1} \leq \alpha \leq \alpha_n . \tag{7.6}$$

The corresponding n mathematical models for these subintervals can be expressed as differential equations:

$$dy/dt = f_1(y, \alpha) , \qquad dy/dt = f_2(y, \alpha) , \quad ... , \quad dy/dt = f_n(y, \alpha) . \tag{7.7}$$

Then, we can define a set of fuzzy if-then rules that basically relate the subintervals to the mathematical models in a one-to-one fashion. The advantage of using fuzzy rules (instead of conventional simple if-then rules) is that we can manage the underlying uncertainty of this process of model selection. We show the set of fuzzy rules for model selection in Table 7.1.

Table 7.1 Decision scheme for model selection

IF		THEN
$\alpha_0 \leq \alpha < \alpha_1$	M_1:	$dy/dt = f_1(y, \alpha)$
$\alpha_1 \leq \alpha < \alpha_2$	M_2:	$dy/dt = f_2(y, \alpha)$
$\alpha_2 \leq \alpha < \alpha_3$	M_3:	$dy/dt = f_3(y, \alpha)$
$\vdots$		$\vdots$
$\alpha_{n-1} \leq \alpha \leq \alpha_n$	M_n:	$dy/dt = f_n(y, \alpha)$

To implement this decision scheme, we need a reasoning method that can use differential equations as consequents. We have developed a new fuzzy inference system that can be considered as a generalization of Sugeno's inference system (Sugeno & Kang, 1988) in which we are now considering differential equations as consequents of the fuzzy rules, instead of simple polynomials. Using this method, the decision scheme of Table 7.1 can be expressed as a single-input fuzzy model as follows:

$$\left\{ \begin{array}{lll} \text{If} & \alpha \text{ is small} & \text{then} \quad dy/dt = f_1(y,\alpha) \\ \text{If} & \alpha \text{ is regular} & \text{then} \quad dy/dt = f_2(y,\alpha) \\ \text{If} & \alpha \text{ is medium} & \text{then} \quad dy/dt = f_3(y,\alpha) \\ & \quad . & \quad\quad\quad . \\ & \quad . & \quad\quad\quad . \\ & \quad . & \quad\quad\quad . \\ \text{If} & \alpha \text{ is large} & \text{then} \quad dy/dt = f_n(y,\alpha) \end{array} \right.$$

where the output y is obtained by the numerical solution of the corresponding differential equation. We have to note here that this new fuzzy inference system reduces to the standard Sugeno system only when the differential equations have closed-form solutions in the form of polynomials. However, the solutions to the differential equations can be more complicated analytical functions or in most cases the solutions are so complex that can only be approximated by numerical methods. The advantage of this generalization of Sugeno's original method is that, in general, we can represent more complicated dynamic behaviors and also because of this fact, the number of rules needed to represent a given dynamical system is smaller.

In Figure 7.3, we show the reasoning procedure for our fuzzy inference system for the case of a one-input single-output fuzzy model. The procedure is very similar to the original Sugeno's procedure, except that now in the output we obtain the crisp values of "y" by solving numerically the corresponding differential equations. The numerical solutions of the differential equations can be achieved by the standard Runge-Kutta type method (Nakamura, 1997):

$$y_{n+1} = RK(y_n) = y_n + 1/2(k_1 + k_2)$$
$$k_1 = hf(y_n, t_n)$$
$$k_2 = hf(y_n + k_1, t_{n+1})$$

where h is the step size of the numerical method and RK can be considered as the Runge-Kutta operator that transforms numerical solutions from time n to time n+1.

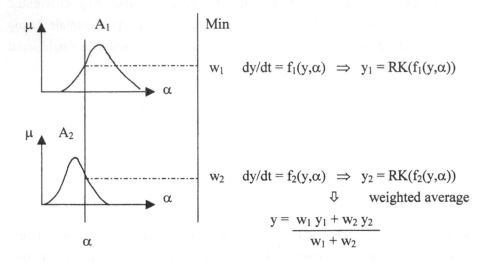

Figure 7.3 The fuzzy inference system for differential equations.

The reasoning procedure for differential equations can also be used for rules with multiple inputs (for the case of several selection parameters) by simply considering the minimum ("min" operator in Figure 7.3) of the firing strengths of each of the inputs. The fuzzy inference system for differential equations can also be illustrated as in Figure 7.4, where a complex dynamical system is modeled by using four different mathematical models (M_1, M_2, M_3 and M_4).

Of course, for this decision scheme to work we need to define membership functions for the different values of the parameter α corresponding to the mathematical models. The membership functions for the models should give us the degree of belief that a particular model is the correct one for a specific value of the parameter α. In Figure 7.5 we show a general method for defining the membership functions for n = 4 models.

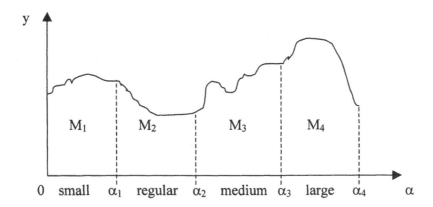

Figure 7.4 Modelling a complex dynamical system with the
fuzzy inference system.

In Figure 7.5, the membership functions for models M_2 and M_3 are of the
"gaussian" type and the membership functions for models M_1 and M_4 are
"sigmoidal". In this way, we can guarantee that there exists a smooth transition
between the degree of membership between the different mathematical models.

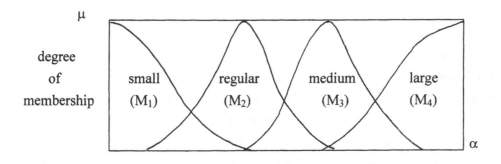

Figure 7.5 General membership functions for n = 4 mathematical models.

To apply this method of model selection, to a particular application, we
have to find the corresponding selection parameter α (or even several parameters)

to be used in the decision scheme proposed in Table 7.1. Then, a partition of the definition interval for α has to be performed. After this, the one-to-one map between the mathematical models and the subintervals (obtained from the partition) is constructed. In this way, we can obtain the fuzzy rule base for model selection for a particular application.

7.4 Neuro-Fuzzy Adaptive Model-Based Control

In this section, we combine the method for adaptive model-based control using neural networks (described in Section 7.2) with the method for model selection using fuzzy logic (described in Section 7.3) to obtain a new hybrid neuro-fuzzy method for control of non-linear dynamical systems. This new method combines the advantages of neural networks (ability for identification and control) with the advantages of fuzzy logic (ability for decision and use of expert knowledge) to achieve the goal of robust adaptive control of non-linear dynamical systems. The general structure of the adaptive system for control is shown in Figure 7.6. In this figure, a module for Model Selection based on fuzzy logic is added to the structure that we had in Figure 7.2, in this way the method can now change between mathematical models according to the dynamic conditions of the plant.

An intelligent control system with the structure shown in Figure 7.6 is capable of adapting to changing dynamic conditions in the plant, because it can change the control actions (given by the neural networks N_c) according to the data that is been measured on-line and also can change the reference mathematical model if there is a large enough change in the value of the selection parameter α. Of course, a change in the reference mathematical model also causes that the neural network N_i performs a new identification for the model. This is the reason why the whole process is called adaptive model-based control of non-linear dynamical systems.

The architecture shown in Figure 7.6 can be used for constructing intelligent control systems for different applications. This can be done by defining the appropriate set of mathematical models for the particular application (according to the type and complexity of the plant or system) and the correct architecture of the neural networks for identification and control. Initial training data can then be used to obtain the initial weights for the networks. The intelligent control system will then be ready for use on-line in the real plant or dynamical system. We have implemented a prototype intelligent control system, with the neuro-fuzzy approach for control, using the MATLAB© programming language. The intelligent system for adaptive model-based control is shown in Appendix C of this book. The computer program listed in this appendix can be used as a basis for developing intelligent control systems for different applications and will be explained in more detail in Chapter 9.

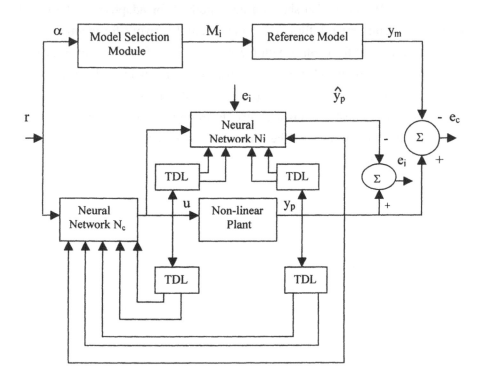

Figure 7.6 Indirect adaptive model-based neuro-fuzzy control.

7.5 Summary

We have presented in this chapter a new method for adaptive model-based control of non-linear dynamical systems. This method is based on a hybrid neuro-fuzzy approach to achieve, in an efficient way, adaptive robust control of non-linear dynamical systems using a set of different mathematical models. We use fuzzy logic to select the appropriate mathematical model for the dynamical system according to the changing conditions of the system. To this end, we have developed a new fuzzy inference system for sets of differential equations, that can be considered a generalization of Sugeno's original method for fuzzy reasoning with polynomials. We have also shown a new method for adaptive model-based control using a neural network for control and a neural network for identification. Combining this method for control with the procedure for fuzzy model selection, gives us a new method for adaptive model-based control using a hybrid neuro-fuzzy approach. This method for adaptive control can be used for general dynamical systems or non-linear plants, since its architecture is independent of the application or the domain and will be illustrated for several complex non-linear problems in Chapter 9.

Chapter 8

Advanced Applications of Automated Mathematical Modelling and Simulation

In this chapter, we present several advanced applications of the new method for automated mathematical modelling and simulation described in Chapters 5 and 6 of this book. First, we describe the application of the new methods for automated modelling and simulation to robotic dynamic systems, which is a very important application in the control of real-world robot arms and general robotic systems. Second, we apply our new methods for modelling and simulation to the problem of understanding the dynamic behavior of biochemical reactors in the food industry, which is also very important for the control of this type of dynamical systems. Third, we consider the problem of modelling and simulation of international trade dynamics, which is an interesting problem in economics and finance. Finally, we also consider the problem of modelling and simulation of aircrafts, as this is important for the real-world problem of automatic aircraft control. We conclude this chapter with some concluding remarks and also some future directions of research work.

8.1 Modelling and Simulation of Robotic Dynamic Systems

Robotic Dynamic Systems can be modelled by systems of coupled non-linear differential equations and then it is possible to have a wide range of possible dynamic behaviors, including the "chaotic" behavior explained above. Of course, this kind of behavior is not desirable in this type of dynamic systems because we need stable robotic systems in the applications. For this reason, it is important to obtain the right mathematical models for the robotic systems and then perform numerical simulations on the models to obtain the information needed in the design and control of these systems. In this section we will review the different methods that can be used to derive mathematical models of robotic systems.

8.1.1 Mathematical modelling of robotic systems

Robot arm dynamics deals with the mathematical formulations of the equations of robot arm motion. The dynamic equations of motion of a manipulator are a set of mathematical equations describing the dynamic behavior of the manipulator (Fu, Gonzalez & Lee, 1987). Such equations of motion are useful for computer simulation of the robot arm motion, the design of suitable control equations for a robot arm, and the evaluation of the kinematic design and structure of a robot arm (Lilly, 1993). The actual dynamic model of a robot arm can be obtained from known physical laws such as the laws of Newtonian mechanics and Lagrangian mechanics. This leads to the development of the dynamic equations of motion for the various articulated joints of the manipulator in terms of specified geometric and inertial parameters of the links. Conventional approaches like the Lagrange-Euler (L-E) and Newton-Euler (N-E) formulations could then be applied systematically to develop the actual robot-arm motion equations. (Fu, Gonzalez & Lee, 1987). These motion equations are "equivalent" to each other in the sense that they describe the dynamic behavior of the same physical robot manipulator. We will consider only the L-E formulation in the following since we are interested in the model of a robotic system in continuous-time.

The derivation of the dynamic model of a manipulator based on the L-E formulation is simple and systematic. Assuming rigid body motion, the resulting equations of motion are a set of second-order coupled non-linear differential equations. The L-E equations of motion provide explicit state equations for robot dynamics and can be utilized to analyze and design advanced joint-variable space control strategies. The derivation of the dynamic equations of an n degrees of freedom manipulator is based on the Lagrange-Euler equation:

$$\frac{d}{dt}\left[\frac{\partial L}{\partial q'_i}\right] - \frac{\partial L}{\partial q_i} = \tau_i \qquad i = 1, 2, ..., n \qquad (8.1)$$

where: L = Lagrangian function = kinetic energy k - potential energy p

  k = total kinetic energy of the robot arm

  p = total potential energy of the robot arm

  q_i = generalized coordinates of the robot arm

  q'_i = first derivative of the generalized coordinate, q_i

  τ_i = generalized force (or torque) applied to the system at joint i to drive
  link i

From the above Lagrange-Euler equation, one is required to properly choose a set of "generalized coordinates" to describe the system. Generalized coordinates are used as a convenient set of coordinates which completely describe the location (position and orientation) of a system with respect to a reference coordinate frame.

Applying the Lagrange-Euler formulation to the Lagrangian function of the robot arm (Fu, Gonzalez & Lee, 1987) yields the necessary generalized torque τ_i for joint i actuator to drive the ith link of the manipulator,

$$\tau_i = \sum_{k=1}^{n} D_{ik}q''_k + \sum_{k=1}^{n}\sum_{m=1}^{n} h_{ikm}q'_kq'_m + c_i \qquad 1 = 1, 2, ..., n \qquad (8.2)$$

or in a matrix form as

$$\tau(t) = D(q(t))q''(t) + h(q(t), q'(t)) + c(q(t)) \qquad (8.3)$$

where $\tau(t)$ = n x 1 generalized torque vector applied at joints i = 1, ..., n ; that is,

$$\tau(t) = (\tau_1(t), \tau_2(t), ... , \tau_n(t))^T \qquad (8.4)$$

  q(t) = an nx1 vector of the joint variables of the robot arm and can be
  expressed as

$$q(t) = (q_1(t), q_2(t), ... , q_n(t))^T \qquad (8.5)$$

  q'(t) = an nx1 vector of the joint velocity of the robot arm and can be expressed as

$$q'(t) = (q'_1(t), q'_2(t), \ldots, q'_n(t))^T \qquad (8.6)$$

$q''(t)$ = an nx1 vector of the acceleration of the joint variables $q(t)$ and can be expressed as

$$q''(t) = (q''_1(t), q''_2(t), \ldots, q''_n(t))^T \qquad (8.7)$$

$D(q)$ = an nxn inertial acceleration-related symmetric matrix

$h(q,q')$ = an nx1 non-linear coriolis and centrifugal force vector whose elements are

$$h(q,q') = (h_1, h_2, \ldots, h_n)^T \qquad (8.8)$$

$c(q)$ = an nx1 gravity loading force vector whose elements are

$$c(q) = (c_1, c_2, \ldots, c_n)^T \qquad (8.9)$$

We will show as an example the Lagrange-Euler equations of motion for a two-link manipulator (Figure 8.1). We assume the following: joint variables = θ_1, θ_2; mass of the links = m_1, m_2; link parameters = $\alpha_1 = \alpha_2 = 0$; $d_1 = d_2 = 0$; and $a_1 = a_2 = 1$. Then, from Equation (8.3) we can obtain that for the two-link manipulator:

$$\tau(t) = D(\theta)\,\theta''(t) + h(\theta, \theta') + c(\theta) \qquad (8.10)$$

where:

$$\tau(t) = \begin{bmatrix} \tau_1 \\ \tau_2 \end{bmatrix}, \qquad \theta''(t) = \begin{bmatrix} \theta''_1 \\ \theta''_2 \end{bmatrix},$$

$$D(\theta) = \begin{bmatrix} (1/3)\,m_1 l^2 + (4/3)\,m_2 l^2 + m_2 c_2 l^2 & (1/3)\,m_2 l^2 + (1/2)\,m_2 l^2 c_2 \\ (1/3)\,m_2 l^2 + (1/2)\,m_2 l^2 c_2 & (1/3)\,m_2 l^2 \end{bmatrix}$$

$$h(\theta, \theta') = \begin{bmatrix} -(1/2)\,m_2 s_2 l^2 \theta'^2_2 - m_2 s_2 l^2 \theta'_1 \theta'_2 \\ (1/2)\,m_2 s_2 l^2 \theta'^2_1 \end{bmatrix}$$

$$c(\theta) = \begin{bmatrix} (1/2)\,m_1 glc_1 + (1/2)\,m_2 glc_{12} + m_2 glc_1 \\ (1/2)\,m_2 glc_{12} \end{bmatrix}$$

As a result of this, we can say that the equations of motion for a two-link manipulator are a set of two second-order coupled non-linear differential equations. By an appropriate change of variables this mathematical model of a two link-manipulator can also be viewed as a set of four first-order coupled non-linear differential equations. Then, the range of dynamic behaviors for (8.10) can go from simple periodic (stable) orbits to even complicated "chaotic" attractors. Of course, more complicated robotic systems will need mathematical models of even higher complexity (in the number of equations and the number of terms) and the identification of dynamic behaviors becomes a real problematic issue. This is the reason why new methods for automated modelling and simulation of robotic dynamic systems are needed and we think that the work presented in this book is a contribution in this line of research.

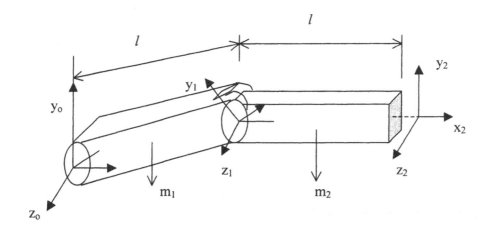

Figure 8.1 A two-link manipulator.

8.1.2 Automated mathematical modelling of robotic dynamic systems

The general method for automated mathematical modelling, described in Chapter 5 of this book, can be used to automate the process of modelling robotic dynamic systems. We only need to specify the set of mathematical models for a specific domain of Robotics and also to define the appropriate values for the variables involved in the process of modelling. We will consider, in this section, the case of modelling robotic manipulators to illustrate the application of the method for automated modelling (Castillo & Melin, 1997b). The general mathematical model for this kind of robotic system is the following:

$$M(q)q'' + V(q, q'))q' + G(q) + F_d q' = \tau \qquad (8.11)$$

where $q \in R^n$ denotes the link position, $M(q) \in R^{nxn}$ is the inertia matrix, $V(q,q')$ $\in R^{nxn}$ is the centripetal-Coriolis matrix, $G(q) \in R^n$ represents the gravity vector, $F_d \in R^{nxn}$ is a diagonal matrix representing the friction term, and τ is the input torque applied to the links.

For the simplest case of a one-link robot arm, we have the scalar equation:

$$M_q q'' + F_d q' + G(q) = \tau \qquad (8.12)$$

If $G(q)$ is a linear function ($G = Nq$), then we have the "linear oscillator" model:

$$q'' + aq' + bq = c$$

where $a = F_d/M_q$, $b = N/M_q$ and $c = \tau/M_q$. This is the simplest mathematical model for a one-link robot arm. More realistic models can be obtained for more complicated functions $G(q)$. For example, if $G(q) = Nq^2$, then we obtain the "quadratic oscillator" model:

$$q'' + aq' + bq^2 = c \qquad (8.13)$$

where a, b and c are defined as above.

A more interesting model is obtained if we define $G(q) = N \sin q$. In this case, the mathematical model is

$$q'' + aq' + b \sin q = c \qquad (8.14)$$

where a, b and c are the same as above. This is the so-called "sinusoidally forced oscillator". More complicated models for a one-link robot arm can be defined similarly.

For the case of a two-link robot arm, we can have two simultaneous differential equations as follows:

$$q''_1 + a_1 q'_1 + b_1 q^2_2 = c_1 \qquad (8.15)$$
$$q''_2 + a_2 q'_2 + b_2 q^2_1 = c_2$$

which is called the "coupled quadratic oscillators" model. In Equation (8.15) a_1, b_1, a_2, b_2, c_1 and c_2 are defined similarly as in the previous models. We can also have the "coupled cubic oscillators" model:

$$q''_1 + a_1q'_1 + b_1q^3_2 = c_1 \qquad\qquad (8.16)$$
$$q''_2 + a_2q'_2 + b_2q^3_1 = c_2$$

We can also have the "coupled forced quadratic oscillators" model:

$$q''_1 + a_1q'_1 + b_1q^2_1 = c_1\sin q_2 \qquad\qquad (8.17)$$
$$q''_2 + a_2q'_2 + b_2q^2_2 = c_2\sin q_2$$

which is a system of two coupled second-order non-linear differential equations. More complicated models for a two-link robot arm can be defined similarly.

Finally, we will consider the case of three-link robot arm. In this case, the mathematical models consist of a set of three simultaneous differential equations of the following form:

$$q''_1 + a_1q'_1 + b_1q_1 = c_1\sin q_2\sin q_3 \qquad\qquad (8.18)$$
$$q''_2 + a_2q'_2 + b_2q_2 = c_2\sin q_1\sin q_3$$
$$q''_3 + a_3q'_3 + b_3q_3 = c_3\sin q_1\sin q_2$$

where the constants are defined in a similar way. This mathematical model can be called "three coupled strongly forced oscillators".

The new method for automated mathematical modelling was defined before (in Chapter 5) for general mathematical models from Dynamical Systems Theory. However, we need to define it now for robotic dynamic systems. This can be accomplish, by making the appropriate changes to the general method described in Chapter 5. We do not have to change the time series analysis module, because the classification scheme for the time series components is valid for any type of dynamical system. We also do not have to change the best model selection module, because the criteria to select the model is still valid. On the other hand, we definitely have to change the expert selection module, because we now have to specify the models appropriate for robotic dynamic systems. We have developed a fuzzy rule base for model selection, for the case of robotic systems, which selects the mathematical models that are the most appropriate with the data available for the given problem. We show in Table 8.1 some sample rules of this knowledge base for model selection.

Table 8.1 Sample fuzzy rules for model selection for robotic systems.

IF			THEN
No. of Links	Trend	Periodic Part	Mathematical Model
1	linear	null	linear_oscillator
1	non-linear	simple	quadratic_oscillator
1	non-linear	regular	cubic_oscillator
1	non-linear	difficult	forced_quadratic_oscillator
1	non-linear	very_difficult	forced_cubic_oscillator
1	non-linear	chaotic	strongly_forced_oscillator
2	linear	null	double_linear_oscillators
2	non-linear	simple	coupled_quadratic_oscillators
2	non-linear	regular	coupled_cubic_oscillator
2	non-linear	difficult	coupled_forced_quadratic_oscillator
2	non-linear	very_difficult	coupled_forced_cubic_oscillators
2	non-linear	chaotic	coupled_strongly_forced_oscillator

The new method for automated mathematical modelling of robotic dynamic systems was implemented in the PROLOG programming language. A prototype intelligent system for automated modelling of robotic manipulator can be found in Appendix A of this book. We have tested the prototype intelligent system with different data to validate the new method and also the implementation with very good results (Castillo & Melin, 1998a). We show below some of the results obtained with the intelligent system for automated modelling of robotic dynamic systems, to give an idea of the performance of the system.

In Figure 8.2 we show the results obtained with the intelligent system for automated modelling of robotic systems for two different cases. First, for a time series with fractal dimension of 0.9 and a robotic system of one link, we can see that the proposed mathematical model is the "linear oscillator" model. The reason why this model is the best one for this case, is because the complexity of the time series is small (which can be classified as "smooth") and the robot arm has only one link.

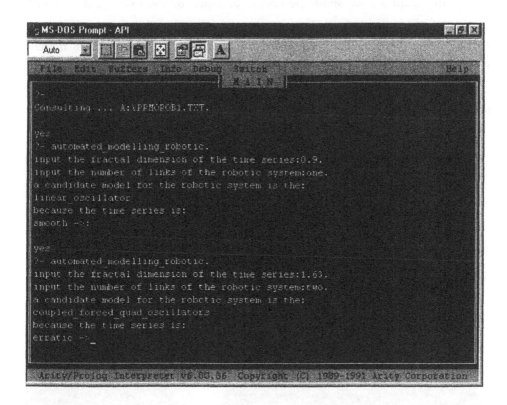

Figure 8.2 Sample input/output using the intelligent system for automated
mathematical modelling of robotic dynamic systems
(First set of cases).

The second case in Figure 8.2 is for a time series with a fractal dimension of 1.63 and two variables, we can see that the proposed mathematical model is the "coupled forced quadratic oscillators". The reason why this model is considered the best one for this case, is because the time series is considered "erratic" (the fractal dimension is relatively high) and the robot arm has two links (in the above cases q is the generalized coordinate of the robot arm).

In Figure 8.3 we show the results obtained by the intelligent system for two more cases. First, for a time series with fractal dimension of 1.35 and a robotic system of one link, we can see that the proposed mathematical model is the "quadratic oscillator". The reason why this model is the best one for this case, is because the complexity of the time series is not that small (which is classified as "erratic" by the program) and the robot arm has only one link.

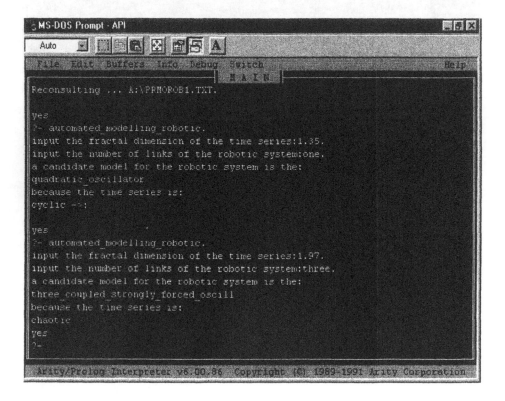

Figure 8.3 Sample input/output using the intelligent system for automated
mathematical modelling of robotic dynamic systems
(Second set of cases).

The second case in Figure 8.3 is for a fractal dimension of 1.97 and a robotic system of three links, in this case the proposed mathematical model is the "three coupled strongly forced oscillators". The reason why this model is considered the best one for this case, is because the time series is considered "chaotic" (the fractal dimension is very high) and the robot arm has three links.

In Figure 8.4 we show the results obtained by the intelligent system for two more cases. First, for a time series with fractal dimension of 1.54 and a robotic system of two links, we can see that the proposed mathematical model is the "coupled cubic oscillators". The reason why this model is considered the best one for this case, is because the complexity of the time series is not that small (which is classified as "erratic" by the program) and the robot arm has two links.

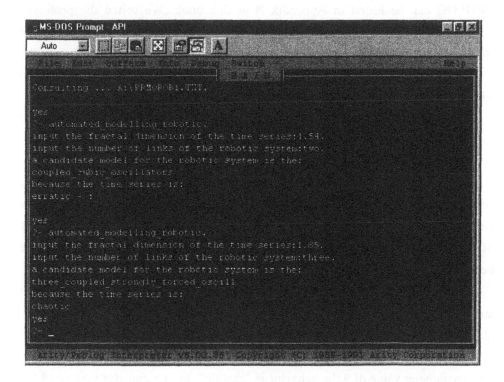

Figure 8.4 Sample input/output using the intelligent system for automated
mathematical modelling of robotic dynamic systems
(Third set of cases).

The second case in Figure 8.4 is for a fractal dimension of 1.85 and a robotic system of three links, in this case the proposed mathematical model is the "three coupled strongly forced oscillators". The reason why this model is considered the best one for this case, is because the time series is considered "chaotic" and the robot arm has three links.

8.1.3 Automated simulation of robotic dynamic systems

Our new method for automated simulation of non-linear dynamical systems was described in Chapter 6 and a prototype implementation of this method in PROLOG can be found in Appendix B of this book. We tested the prototype intelligent system with different data to validate the new method and also the implementation with very good results. In this section, we show some of the results obtained using the intelligent system for automated simulation, to give an idea of the performance of the system.

In Figure 8.5 we show the results obtained for a random initial population (of three members) and a simple mathematical model (given as a set of facts in the program of Appendix B). We can see in Figure 8.5 how the genetic algorithm evolves the initial population in such a way that three different dynamical behaviors are identified for the three corresponding parameter values. For a parameter value of 3 the behavior is a "cycle of period two", for a parameter value of 14 the behavior is a "cycle of period eight", and for a parameter value of 4 the behavior is a "cycle of period four".

In Figure 8.6 we show the results obtained for a random initial population and a simple mathematical model. We can see in Figure 8.6 how the genetic algorithm evolves the initial population in such a way that three different dynamical behaviors are identified for the three corresponding parameter values. For a parameter value of 9 the behavior is "chaotic", for a parameter value of 4 the behavior is a "cycle of period four", and for a parameter value of 1 the behavior is a "fixed point".

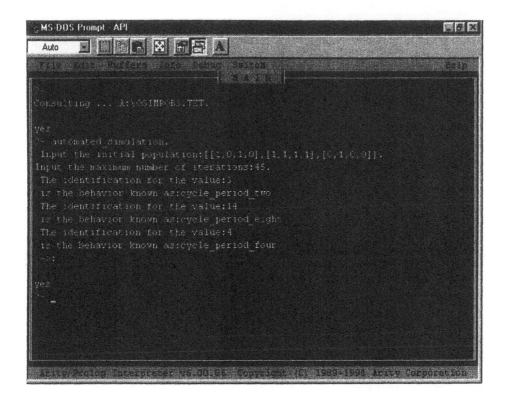

Figure 8.5 Sample input/output using the intelligent system for automated
simulation with an initial population of [10, 15, 4].

We also show in this section some simulation results obtained with a
prototype intelligent system for automated simulation that was developed in
MATLAB, for several types of robotic dynamic systems. The prototype intelligent
system in the MATLAB programming language runs more efficiently because
MATLAB can perform numerical calculations faster than PROLOG. Also,
MATLAB offers advantages in visualizing the results graphically. The numerical
simulation results are briefly explained to give an idea of the performance of the
intelligent system.

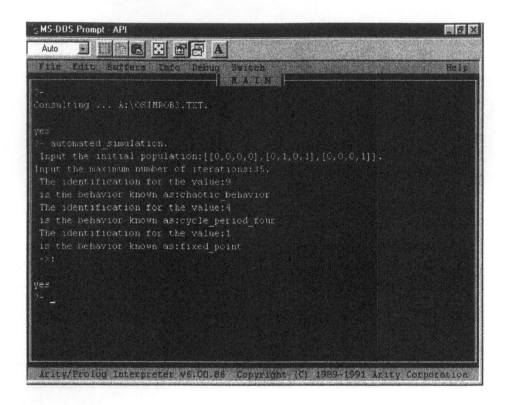

Figure 8.6 Sample input/output using the intelligent system for automated
simulation with an initial population of [0, 5, 1].

In Figure 8.7 we show the simulation results for a single-link robotic
dynamic system with a mathematical model given by the second order differential
equation:

$$q'' + aq' + b\sin q = c \qquad\qquad (8.19)$$

where a, b and c are physical parameters of the robotic system. The simulation
results shown in Figure 8.7 correspond to the parameters: a = 30, b = 60, c = 7
and to the initial conditions: q(0) = 5, q'(0) = 5. The solution shown in Figure 8.7
is what is known as a cycle of period two because the orbit is oscillating between

two different points. As a consequence of this the behavior identification in this case is of a "cycle of period two".

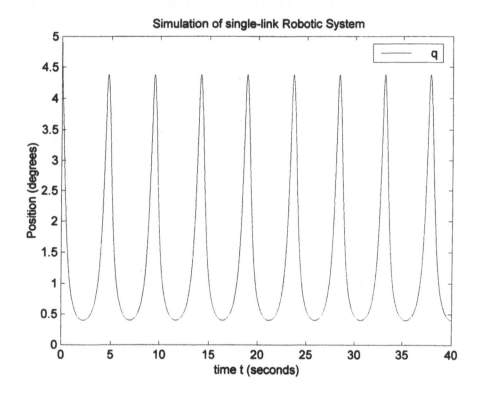

Figure 8.7 Simulation of a single-link robot arm showing a cycle of period two.

In Figure 8.8 we show the simulation results for a single-link robotic dynamic system with a mathematical model given by Equation (8.19) The parameters remain the same as in the previous case, except for "a" which changes to: a=2. The initial conditions are different to better appreciate the dynamical behavior and are given by: q(0) = 35, q'(0) = 35. The solution shown in Figure 8.8 is what is known as a cycle of period eight because the orbit is oscillating between eight different points (after a transient period). As a consequence of this fact the behavior identification in this case is of a "cycle of period eight".

The explanation for the change of dynamical behavior between a = 30 and a = 2 is related to the "damping" (given by "a") of the forced oscillator given by Equation (8.19). It is a well known fact that less damping implies more oscillatory power for a mechanical system (in this case, the robotic system).

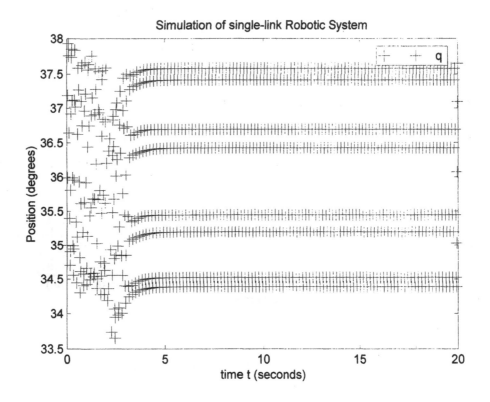

Figure 8.8 Simulation of a single-link robot arm showing a cycle of period eight.

In Figure 8.9 we show the simulation results for a two-link robotic dynamic system with a mathematical model given by the two coupled second order differential equations:

$$q''_1 + a_1q'_1 + b_1\sin q_2 = c_1$$
$$q''_2 + a_2q'_2 + b_2\sin q_2 = c_2 \qquad\qquad (8.20)$$

where a_1, a_2, b_1, b_2, c_1 and c_2 are physical parameters of the robotic system. The simulation results shown in Figure 8.9 correspond to the parameter values:

$$a_1 = 212, \quad a_2 = 44, \quad b_1 = b_2 = 60, \quad c_1 = \quad c_2 = 72$$

and to the initial conditions: $\quad q_1(0) = 0.5, \quad q'_1(0) = 5, \quad q_2(0) = 4, \quad q'_2(0) = 5$.

The Solutions shown in Figure 8.9 are known as cycles of period two because their orbits are oscillating between two different points. The behavior identification in this case is of a "cycle of period two" for both links.

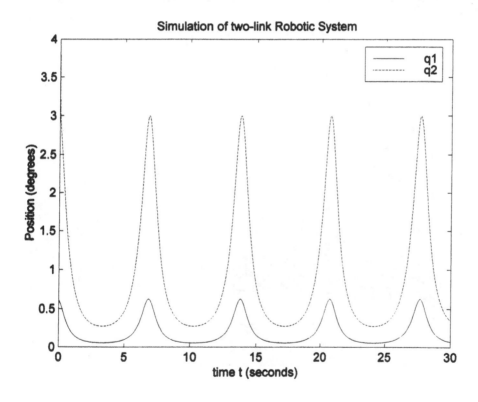

Figure 8.9 Simulation of a two-link robot arm showing cycles of period two for the positions q_1 and q_2 of the links.

In Figure 8.10 we show the simulation results for a two-link robotic dynamic system with a mathematical model given by Equation (8.20). The parameters remain the same as in the previous case, except for "a_2" which changes to: $a_2 = 2$. The initial conditions are different to better appreciate the complex dynamical behavior and are now given by: $q_1(0) = 0.5$, $q'_1(0) = 5$, $q_2(0) = 35$, $q'_2(0) = 5$. The solution shown in Figure 8.10 is what is known as a "chaotic solution" because the orbit is oscillating (in an unstable manner) between an infinite number of periodic points. As a consequence of this fact the behavior identification in this case is of a "chaotic solution".

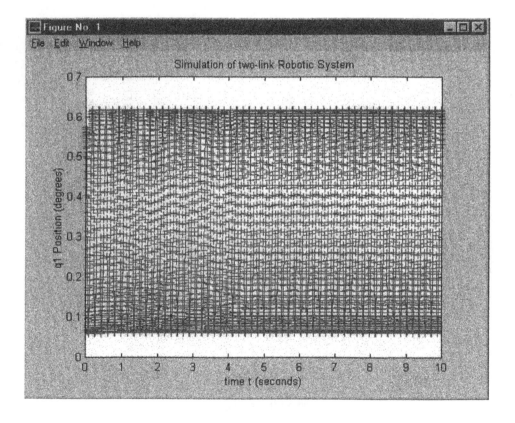

Figure 8.10 Simulation of a two-link robot arm showing chaotic behavior for position q_1.

The explanation for the change of dynamical behavior between the previous case for $a_2 = 44$ and the case for $a_2 = 2$ is similar to the one given before for a single-link robot, the only difference is that now the behavior is more complex (no periodic solution is found).

In Figure 8.11 we also show the simulation results for a two-link robotic system with model given by Equation (8.20). The parameters and initial conditions remain the same as in the last case, except that now q_2 is shown in Figure 8.11. We also have chaotic behavior for position q_2, as in the previous figure, the only difference is in the range of numerical values of q_2.

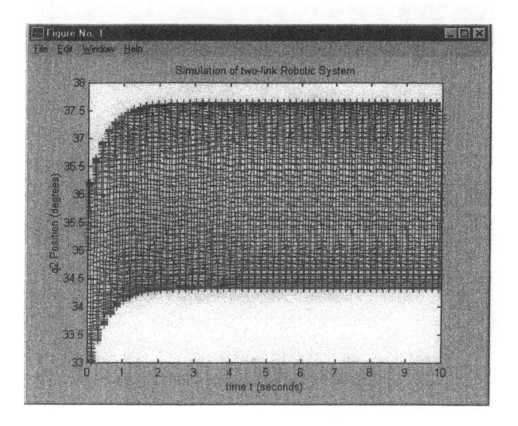

Figure 8.11 Simulation of a two-link robot arm showing chaotic behavior for position q_2.

Finally, in Figure 8.12 we show the simulation results for the two-link robotic system with the same model and parameters as in the last two cases, except that now q_2 and q_1 are shown in the same figure. In this figure, we can see how both q_1 and q_2 tend to a "strange attractor", which is one of the distinguishing signs of "chaotic" behavior (Rasband, 1990). Of course, in Robotic applications this behavior has to be avoided because it will cause physical damage to the robotic system. This is why it is important to identify when this behavior can occur in advance to avoid critical situations.

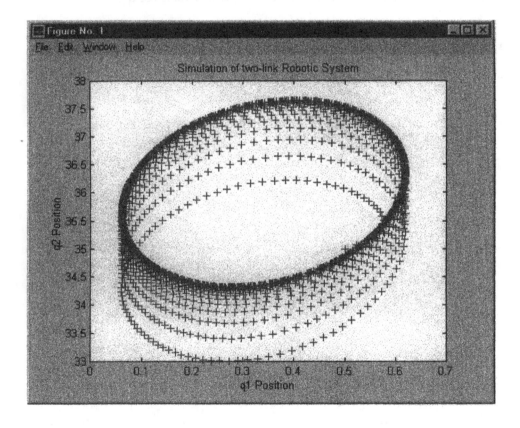

Figure 8.12 Simulation of a two-link robot arm showing chaotic behavior for positions q_1 and q_2.

8.2 Modelling and Simulation of Biochemical Reactors

We describe in this section the application of the new methods for modelling and simulation (described in Chapters 5 and 6) to the complex case of biochemical reactors. The case of biochemical reactors for food processing plants is a very complex one, because biochemical processes are often highly non-linear and consequently difficult to control (Melin & Castillo, 1998a). First, we consider the problem of modelling biochemical reactors using non-linear differential equations. We then describe how to automate this process of modelling, using the fuzzy-fractal approach described in Chapter 5. We also describe how to automate the process of numerical simulation for the mathematical models, using the fuzzy-genetic approach described in Chapter 6. Finally, we show some simulation results for the mathematical models of the biochemical reactors.

8.2.1 Modelling biochemical rectors in the food industry

Many products of considerable economic value of us humans are the result of the metabolic functions of microorganisms. From the industrial point of view, the substrate can be considered as the raw material and the microorganisms as the biochemical "micro-industry" that transforms this material into new products. In this section, we consider briefly the problem of food production using bacteria for industrial purposes. Many different food products are the result of bacteria population and the best example of this is the set of dairy (milk) products (Pelczar & Reid, 1982).

We will consider the problem of food production for the case of yogurt. In this case, the use of Streptococcus thermophilus and Lactobacillus bulgaricus is to produce Lactic Acid which is the critical chemical compound necessary for obtaining this product with the exact biochemical properties. The right use of this two types of bacteria in this case, i.e. right temperature and time, results in more quantity and better quality of yogurt. We can say then, that from the industrial point of view the goal is to obtain the maximum quantity of the food product with the desired chemical and biological properties (Melin & Castillo, 1997d).

However, this goal from the engineering point of view can be translated to obtaining the "best" control possible for the food generation process, and this involves modelling bacteria population in the substrate.

For the case of yogurt we need to have mathematical models for the population of S. thermophilus and L. bulgaricus and the quantity of Lactic Acid produced by both bacteria. In this case, we require a Lotka-Volterra type model consisting of a system of three simultaneous differential equations, modelling the situation of two bacteria populations and one chemical compound concentration (Melin & Castillo, 1996).

We will consider first the case of using only one bacteria for food production. The mathematical model in this case can be of the following form:

$$dN/dt = r(1-N/K)N - \beta N \qquad (8.21)$$
$$dP/dt = \beta N$$

where: N = population of bacteria, P = quantity of chemical product, r = rate of growth of the bacteria, k = limiting capacity of the environment (substrate quantity) and β = biochemical conversion factor. We can see how Equation (8.21) is of the general form given by the mathematical model of a non-linear plant of a single input and single output (Equation (7.1) of Chapter 7).

We will consider now the case of two bacteria used for food production:

$$dN_1/dt = [r_1 - (r_1/K_1)N_1 - (r_1/K_1)\alpha_{12}N_2]N_1 - \beta N_1 \qquad (8.22)$$
$$dN_2/dt = [r_2 - (r_2/K_2)N_2 - (r_2/K_2)\alpha_{21}N_1]N_2 - \gamma N_2$$
$$dP/dt = \beta N_1 + \gamma N_2$$

where: N_1 = population of bacteria 1, N_2 = population of bacteria 2, P = quantity of chemical product, r_1 = rate of growth of bacteria 1, r_2 = rate of growth of bacteria 2, K_1=capacity of the environment for bacteria 1, K_2=capacity of the environment for bacteria 2, β=biochemical conversion factor from bacteria 1 to product, γ = biochemical conversion factor from bacteria 2 to product, α_{12} and α_{21} are coefficients of the system. We can see how Equation (8.22) is of the general form given by the model of a non-linear plant of Equation (7.2) of Chapter 7.

Another interesting case will be if one considers two "good" bacteria for food production and one "bad" bacteria that is "attacking" the other bacteria:

$$dN_1/dt = [r_1 - (r_1/K_1)N_1 - (r_1/K_1)\alpha_{12}N_2]N_1 - \beta N_1 - \sigma_1 N_1 N_3 \qquad (8.23)$$
$$dN_2/dt = [r_2 - (r_2/K_2)N_2 - (r_2/K_2)\alpha_{21}N_1]N_2 - \gamma N_2 - \sigma_2 N_2 N_3$$
$$dN_3/dt = [r_3 - (r_3/K_3)N_3] N_3 + \sigma_1 N_1 N_3 + \sigma_2 N_2 N_3$$
$$dP/dt = \beta N_1 + \gamma N_2$$

where: N_3 = population of bacteria 3 (bad bacteria), r_3 = rate of growth of bacteria 3, K_3= capacity of the environment for bacteria 3 and σ_1 = rate of attack of bacteria 3 to bacteria 1, and σ_2 = rate of attack of bacteria 3 to bacteria 2. We can see how Equation (8.23) is of the general form given by the mathematical model of a non-linear plant of Equation (7.3) of Chapter 7.

This last two cases are more interesting from the mathematical point of view, because it is a well known fact from Dynamical Systems theory that a model of three or more coupled non-linear differential equations can exhibit the behavior known as "chaos" (Kapitaniak, 1996). This chaotic behavior has to be avoided for this kind of problems, because we need to have a stable food production process. Then, part of the problem in this case will be to control the food production process avoiding at the same time the chaotic regime of this type of models.

In Equations (8.21) - (8.23) we are modelling bacteria growth only in the time domain. If we want also to consider bacteria growth in the space domain, we need to consider a measure of the geometrical complexity of bacteria colonies. A method that classifies bacteria colonies using the fractal dimension was developed for identification purposes (Castillo & Melin, 1994a). This method uses the fractal dimension to make a unique classification of the different types of bacteria, because it is a well known experimental fact that colonies of different types of bacteria have different geometrical forms. Then a one-to-one map can be constructed that relates each type of bacteria to its corresponding fractal dimension.

Now we propose mathematical models that integrate the method for geometrical modelling of bacteria growth (using the fractal dimension) with the method for modelling the dynamics of bacteria population (using differential equations). The resulting mathematical models describe bacteria growth in space and in time, because the use of the fractal dimension enables us to classify bacteria by the geometry of the colonies and the differential equations help us to understand the evolution in time of bacteria population. The models will be similar to the ones described before, except that now the fractal dimension D is integrated into the differential equations (Melin & Castillo, 1997b).

We will consider first the case of using one bacteria for food production. The mathematical model in this case can be of the following form:

$$dN/dt = r(1 - N^{-D}/K)N^{-D} - \beta N^{-D} \tag{8.24}$$

$$dP/dt = \beta N^{-D}$$

where D is the fractal dimension and the rest of the variables are as described before.

We will consider now the case of two bacteria used for food production:

$$dN_1/dt = [r_1 - (r_1/K_1)N_1^{-D1} - (r_1/K_1)\alpha_{12}N_2^{-D2}]N_1^{-D1} - \beta N_1^{-D1} \tag{8.25}$$

$$dN_2/dt = [r_2 - (r_2/K_2)N_2^{-D2} - (r_2/K_2)\alpha_{21}N_1^{-D1}]N_2^{-D2} - \gamma N_2^{-D2}$$

$$dP/dt = \beta N_1^{-D1} + \gamma N_2^{-D2}$$

where D_1 = fractal dimension of bacteria 1, D_2 = fractal dimension of bacteria 2 and the rest of variables are as described before.

We can also propose an equation similar to the one described before (Eq. 8.23) for two "good" bacteria and one "bad" bacteria by using the fractal dimensions, D_1 D_2 and D_3 , for the corresponding types of bacteria. Also, we can apply this method of modelling to more complicated cases of food production.

As we can see from Equations (8.24) and (8.25) the idea of our method of modelling is to use the fractal dimension D as a parameter in the differential equations, so as to have a way of classifying for which type of bacteria the equation corresponds. In this way, Equation (8.24), for example, can represent the model for food production using one bacteria (the one defined by the fractal dimension D). Since, the fractal dimension gives us a unique way to classify bacteria, then also Equation (8.24) gives us a unique way to model the corresponding problem of food production using one bacteria.

8.2.2 Automated mathematical modelling of biochemical reactors

The method for automated modelling using a fuzzy-fractal approach (described in Chapter 5) can be used to select the appropriate mathematical models for biochemical reactors. We only need to make the necessary changes to the rules for model selection. No changes are needed for the "time series analysis" module or the "best model selection" module. The main changes needed to the "Expert Modelling" module are the following:

     1) Use of mathematical models for biochemical reactors
     2) Use of the fractal dimension as a classification variable.

With these changes the general method for automated modelling of dynamical systems (described in Chapter 5) can be transformed to a method for modelling biochemical reactors. We show in Table 8.2 some sample rules of the knowledge for model selection.

     In Table 8.2, the fuzzy value "small" means near to 1, because when the fractal dimension is near to a value of 1, we can use equations (8.21), (8.22) and (8.23). On the other hand, the fuzzy value "large" means greater than 1, because when the fractal dimension is sufficiently different from 1, we have to use equations (8.24) and (8.25). Of course, we have to define the appropriate membership functions for the values "small" and "large" in this fuzzy rule base, to make this method work for the domain of modelling biochemical reactors.

Table 8.2 Sample fuzzy rules for model selection for biochemical reactors.

	IF			THEN
No. of bacteria	Fractal Dimension of bacteria 1	Fractal Dimension of bacteria 2	Fractal Dimension of bacteria 3	Mathematical Model
1	small			Lotka-Volterra of Eq. (8.21)
1	Large			Lotka-Volterra with fractal dimension of Eq. (8.24)
2	small	small		Equation (8.22)
2	Large	small		Equation (8.25)
2	small	Large		Equation (8.25)
2	Large	Large		Equation (8.25)
3	small	small	small	Equation (8.23)
3	Large	small	small	Equation (8.23) with fractal dimension
3	Large	Large	small	Equation (8.23) with fractal dimension
3	Large	Large	Large	Equation (8.23) with fractal dimension

8.2.3 Simulation results for biochemical reactors

We describe in this section the simulation results obtained with the implementation of numerical methods for the approximate solution of differential equations. The complete listings of the computer programs (written in the MATLAB language) can be found in Appendix B. In all cases we used Runge-Kutta type methods to approximate the numerical solution of the mathematical models for the plants. The parameter values of the mathematical models are for the case of real biochemical reactors in the food industry. The numerical simulation of the models (M_1, M_2 and M_3) show the complex dynamics involved in the biochemical process of production. In the following figures, we show some of these results to give an idea of the complexity of biochemical reactors (and as a consequence of the food processing plants) and the degree of difficulty in controlling them.

We show in Figure 8.13 the complicated dynamics involved even for the simplest case of only one bacteria used for food production (for example, yogurt). The mathematical model (M_1) is as given by Equation (8.21).

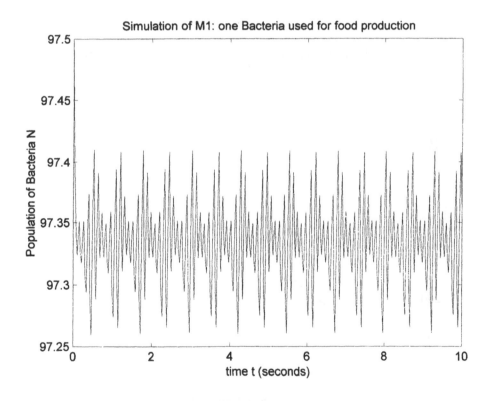

Figure 8.13 Numerical Simulation of the population in the model of one bacteria used for food production.

The numerical simulation shown in Figure 8.13 is for the following initial conditions: initial population: N = 97.5 for t =0   initial product: P = 0   for t = 0 and the parameter values given as follows:

rate of growth of the bacteria: r = 30

limiting capacity of the environment: k = 100

biochemical conversion factor: β = 0.8

In this case, bacteria grows at a rate of 30 % and only 80 % of microbial life is converted into the chemical that produces food. The parameter values are standard in food production and the simulation shows that the dynamics is complicated. We also show in Figure 8.14 the dynamics for the food product for the same case of only one bacteria used for production.

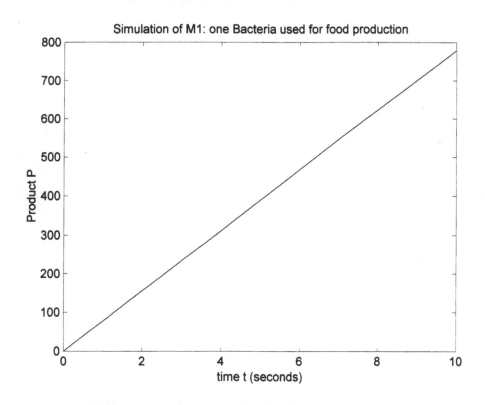

Figure 8.14 Numerical Simulation of the product in the model of one bacteria used for food production.

We can see in Figure 8.14 that the production is linear in time, with a final value of 70% at a time of 10. The mathematical model (M_1) used in the simulation is given by Equation (8.21). The initial conditions and the parameter values are the same as the ones used in Figure 8.13.

    We show in Figure 8.15, the complicated dynamics for the case of two bacteria used for food production (for example, yogurt). The mathematical model (M_2) is given by Equation (8.22).

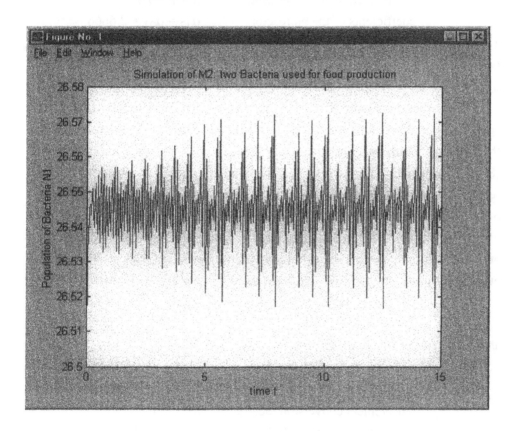

Figure 8.15 Numerical simulation of the population in the model of
two bacteria used for food production.

The numerical simulation shown in Figure 8.15 is for the following initial conditions:

initial populations: $N_1 = N_2 = 26.5$ for $t = 0$

initial product $P = 0$ for $t = 0$

and the parameter values are given as follows:

rate of growth for the bacteria: $r_1 = r_2 = 30$

limiting capacity of the environment: $k_1 = k_2 = 100$

biochemical conversion factors: $\beta = \gamma = 0.8$

coefficients of the bioreactor: $\alpha_{12} = \alpha_{21} = 2.6666$

These parameter values can be interpreted similarly to the case of only one bacteria. These values are standard in food production and show the complicated dynamics of the population for the two bacteria.

We show in Figure 8.16, the complicated dynamics for the case of two "good" bacteria used for food production and one "bad" bacteria that is reducing the efficiency of the process. The mathematical model (M_3) is given by Equation (8.23). We can see in Figure 8.16 how the population N_2 of the "good" bacteria 2 reduces to zero, and how the populations of the "bad" bacteria (N_3) and of the other "good" bacteria (N_1) stabilize the value to a fixed population. The net result of this situation is that the final quantity of the food product decreases because only one bacteria is producing and the other ones are not producing. The numerical simulation in Figure 8.16 is for the following initial conditions:

initial populations: $N_1 = 65$ $N_2 = 6.5$ $N_3 = 10$ for $t = 0$

initial product $P = 0$ for $t = 0$

and the parameter values are given as follows:

rate of growth for the bacteria: $r_1 = r_2 = r_3 = 30$

limiting capacity of the environment: $k_1 = k_2 = k_3 = 100$

biochemical conversion factors: $\beta = \gamma = 0.8$

rate of attack of bacteria 3 $\sigma_1 = \sigma_2 = 0.2$

coefficients of the bioreactor: $\alpha_{12} = \alpha_{21} = 2.6666$

Of course, this is a case we want to avoid in the food production process. In our method, by controlling the temperature we can control the population of the 'bad" bacteria to avoid the corresponding reduction in the production.

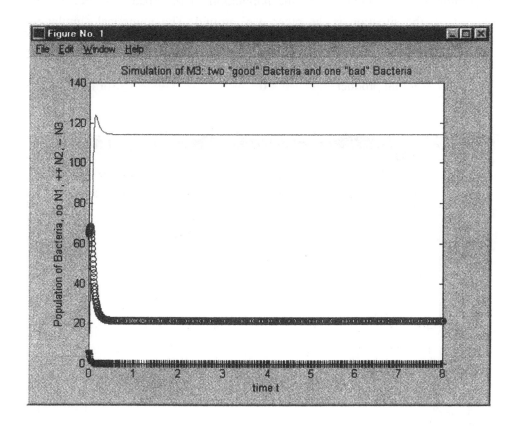

Figure 8.16 Numerical simulation of the populations in the model of three
bacteria in the food production process (case "a").

We show in Figure 8.17, the complicated dynamics for the case of two
"good" bacteria used for production and one "bad" bacteria attacking the other
ones. The mathematical model is the same as in the last case. We can see in
Figure 8.17 how the population N_1 of the "good" bacteria 1 reduces to zero, and
how the other two populations stabilize to a fixed population value. The net result
of this situation is that the final production decreases because only one bacteria is
producing. In this case, the production is higher than the one of case "a" because

the population N_2 is greater than the population of the "bad" bacteria N_3. In Figure 8.16, the situation is the other way around. The numerical simulation shown in Figure 8.17 is for the following initial conditions:

initial populations: $N_1 = N_2 = 60$ $N_3 = 0.5$ for $t = 0$

initial product $P = 0$ for $t = 0$

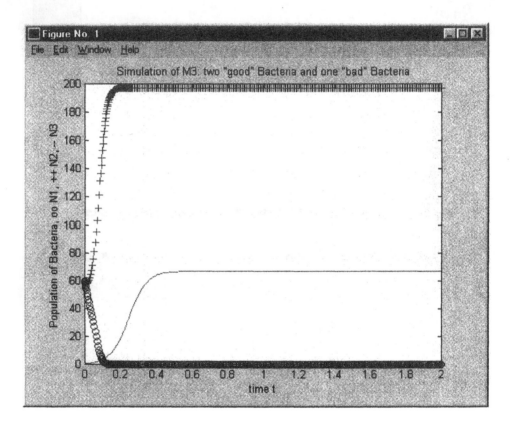

Figure 8.17 Numerical simulation of the populations in the model of three bacteria in the food production process (case "b").

and the parameter values are given as follows:

rate of growth for the bacteria: $r_1 = 50$ $r_2 = 60$ $r_3 = 20$

limiting capacity of the environment: $k_1 = 166.666$ $k_2 = 200$ $k_3 = 66.666$

biochemical conversion factors: $\beta = \gamma = 0.8$

rate of attack of bacteria 3 $\sigma_1 = \sigma_2 = 0.0002$

coefficients of the bioreactor: $\alpha_{12} = \alpha_{21} = 2.6666$

Of course, this is another case we want to avoid in the food production process. The new method for adaptive control will reduce the temperature (or increase it) to reduce the population of the "bad" bacteria and increase production.

8.3 Modelling and Simulation of International Trade Dynamics

We describe in this section mathematical models that can be used to study the dynamics of international trade (Castillo & Melin, 1998c). Mathematical models of International Trade (IT), between three or more countries, can show very complicated dynamics in time (with the possible occurrence of chaotic behavior). The simulation of these models is critical in understanding the behavior of the relevant financial and economical variables for the problem of IT. Also, performing the simulations for different parameter values of the models will enable the forecasting of future IT. The problem of modelling and simulation of IT has been solved in this section by applying the methods described in Chapters 5 and 6 of this book.

8.3.1 Mathematical modelling of international trade

Mathematical modelling of international trade has been done traditionally with linear statistical models from classical Econometric Theory. However, more recently some researchers have found statistical evidence that time series from financial and economical variables show erratic fluctuations in time. It is well known that simple linear models can not represent this erratic dynamic behavior. for this reason, it becomes necessary to use non-linear mathematical models that will enable us to represent this complex dynamic behavior found for systems in economics and finance. Non-linear models from the theory of dynamical systems can show the behavior known as "chaos" for different ranges of parameter values and for this reason they become a good choice in modelling complex financial or economic problems (Castillo & Melin, 1998c).

We will consider first modelling the dynamics of autonomous economies, i.e.., study the oscillations of an autonomous economy. Then, we will consider modelling the problem of International Trade as a perturbation of the internal oscillations of an autonomous economy.

Consider the Keynesian macroeconomic model of a single economy with Y as income, r as the interest rate, M as the (constant) nominal money supply, and assume that the good prices, P, are fixed during the relevant time interval. Suppose that gross investment, I, and savings, S, depend both on income and the interest rate in the familiar way, i.e.,

$$I = I (Y,r) \quad , \qquad\qquad Iy > 0, Ir < 0$$
$$S = S (Y,r) \quad , \qquad\qquad Sy > 0, Sr < 0$$

Income adjusts according to excess demand in the goods market, i.e.,

$$Y' = \alpha (I - S) \qquad\qquad \alpha > 0 \qquad\qquad (8.26)$$

The set of points $\{(Y,r)| I(Y,r) = S(Y,r)\}$ constitutes the IS-curve of the model. Let $L(Y,r)$ denote the liquidity preference with $Ly > 0, Lr < 0$ and assume that the interest rate adjusts according to:

$$r' = \beta (L(Y,r) - M/p) \qquad\qquad , \beta > 0 \qquad\qquad (8.27)$$

with the set of points $\{(Y,r)| L(Y,r) = M/p\}$ forming the LM-curve of the model. As is well known, the equilibrium (Y^*,r^*) is asymptotically (locally) stable if trJ < 0 and det J >0, where J is the Jacobian of the system and Tr = trace, det = determinant. Also, it can be demonstrated by means of the Poincaré-Bendixon Theorem that system (8.26) - (8.27) is able to generate oscillating behavior.

Consider three economies, each of which is described by equations like (8.26) - (8.27) with possibly different numerical specifications of the functions, i.e.,

$$Y'_i = \alpha_i (I_i(Y_i,r_i) - S_i(Y_i,r_i))$$
$$\qquad\qquad\qquad\qquad\qquad i = 1, 2, 3 \qquad\qquad (8.28)$$
$$r'_i = \beta i (L_i(Y_i,r_i) - M_i/p_i)$$

Equation (8.28) constitutes a six-dimensional differential equation system which can also be written as a system of three independent two-dimensional limit cycle oscillators.

By introducing international trade with linear functions $EX_i = EX_i(Y_j, Y_k)$, $i \neq j, k$ and $Im_i = Im_i(Y_i)$, Equation (8.28) becomes:

$$Y'_i = \alpha_i (I_i(Y_i, r_i) - S_i(Y_i, r_i) + EX_i(Y_j, Y_k) - Im_i(Y_i)) \qquad (8.29)$$
$$r'_i = \beta_i (L_i(Y_i, r_i) - M_i/p_i)$$

with i, j, k = 1, 2, 3; j, k = i, and Mi as the money supplies reflecting balance of payments equilibria. Equation (8.29) constitutes a system of three linearly coupled limit cycle oscillators. The following theorem can then be demonstrated for system (8.29).

Theorem: If all three autonomous economies are oscillating, the introduction of international trade may imply the existence of a strange attractor (chaotic behavior).

Of course, chaotic behavior may occur for certain ranges of parameter values for the α_i, β_i, M_i parameters. However, the emergence of strange attractors is not exclusive in models like these: some variations in the parameters can lead to the occurrence of other phenomena like quasi-periodic motion or phase-locking. The main goal for a certain country is to achieve a stable behavior in its economy while in this International Trade System, in this way controlling its future behavior in this system. As a result of this, a specific country (like Mexico) can optimize its profit while in a system of three countries (like with the NAFTA trade agreement).

The general method for automated mathematical modelling, described in Chapter 5, can be used to automate the process of modelling the problem of international trade. We only need to specify the set of mathematical models (satisfying the general form of Equation (8.29)) and also to define the appropriate values for the variables involved in the process of modelling.

8.3.2 Simulation results of international trade

We show in this section some simulation results obtained using the method for automated simulation of dynamical systems (described in Chapter 6). The fuzzy-genetic approach for simulation enables the automated generation of parameter values for the models and obtains the corresponding identifications of the dynamic behaviors. We will show here only simulation results for the case of three countries and leave to the reader the simulation of other cases.

For the case of USA, Canada and Mexico we can assume that the system of differential equations given by Equation (8.29) represents a good general model of the actual system of the three corresponding economies. Then, using investment and saving functions of the form:

$$I_i = a_i \, Y_i / r_i$$
$$\qquad\qquad\qquad\qquad i = 1,2,3 \qquad\qquad\qquad (8.30)$$
$$S_i = b_i \, Y_i / r_i$$

and with Liquidity given by a similar function:

$$L_i = c_i \, Y_i / r_i \qquad\qquad\qquad i = 1,2,3 \qquad\qquad\qquad (8.31)$$

and assuming that the export and import functions are linear:

$$EX_i = d_i Y_j + e_i Y_k \qquad\qquad i \neq j, k \qquad\qquad\qquad (8.32)$$
$$Im_i = f_i + g_i Y_i \qquad\qquad\qquad i = 1,2,3 \qquad\qquad\qquad (8.33)$$

we can find that a specific mathematical model for the three economies is given by the following system:

$$Y_i' = \alpha_i \, (\, (a_i - b_i) \, (Y_i / r_i) + d_i Y_j + e_i Y_k - f_i - g_i Y_i)$$
$$\qquad\qquad\qquad\qquad\qquad i = 1,2,3 \qquad\qquad\qquad (8.34)$$
$$r_i' = \beta_i \, (\, c_i \, (Y_i / r_i) - M_i / p_i)$$

where α, β, a, b, c, d, e, f, g are parameters to be estimated using actual data of the problem (time series for Y and r). Of course, Equation (8.34) is only one of the specific models that can be explored for this particular problem.

Whether or not there are indeed "strange" attractors and hence chaotic trajectories in a specific model can be established only by numerical techniques. Simulation results for the system of Equation (8.34) indicate that economically reasonable specifications for this model can be found which indeed imply positive Lyapunov exponents and chaos. For this reason, we think that erratic fluctuations in the economy of a specific country can occur when a transition is made from a closed economy to an open one.

We show in Figure 8.18 the dynamic behavior for the investment in the international trade system of Equation (8.34). The parameter values for this simulation were obtained automatically by the intelligent system for automated simulation. We can see in Figure 8.18 how the investment for USA and Canada are growing more rapidly than the one for Mexico in this specific case.

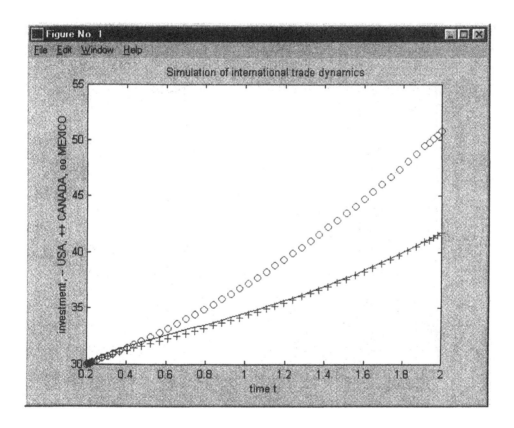

Figure 8.18 Simulation results showing the dynamic behavior for the investment of USA, Canada and Mexico.

The simulation results shown in Figure 8.18 are for a situation of stable growing economies for the three countries. In the present time, the dynamic evolution for the countries (specially for Mexico) is not stable at all, but the ultimate goal is to achieve a stable growing pattern for the international trade system of USA, Canada and Mexico so as to optimize the profits for the three countries. The situation shown in Figure 8.18 could be a goal state for the three economies.

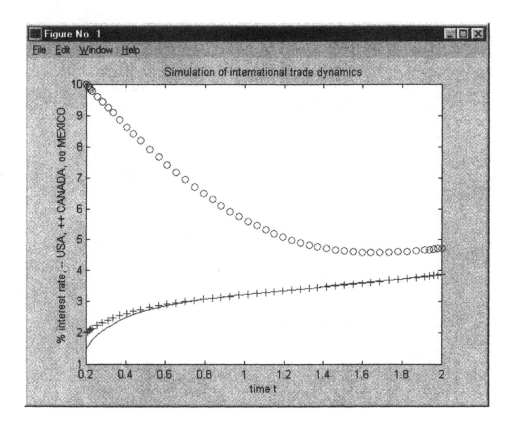

Figure 8.19 Simulation results showing the dynamic behavior for the interest rate of USA, Canada and Mexico.

We show in Figure 8.19 the simulation results for the interest rate dynamic evolution in time, for the same case of stable growing economies for the three countries. We can see in Figure 8.19 how the interest rates of the three countries are relatively stable (Mexico's interest rate is higher but can be considered good). Still, we think that more research has to be done regarding which is to be considered the best situation for the three economies in this international trade system.

8.4 Modelling and Simulation of Aircraft Dynamic Systems

We describe in this section mathematical models that can be used to study the dynamics of aircraft systems. Mathematical models of aircraft systems can show very complicated dynamics in time (with possible occurrence of chaos). The simulation of these models is critical in understanding the dynamic behavior of a real airplane. Also, performing the simulations for different parameter values of the models will enable the forecasting of future behavior of the airplane, to avoid possible failures of the aircraft. We solved the problem of modelling and simulation of aircraft systems by applying the methods described in Chapters 5 and 6.

8.4.1 Mathematical modelling of aircraft systems

We now present some simplified mathematical models of aircraft systems to study the dynamics of the aircraft during flight (Melin & Castillo, 1998c). The models are in the form of equations of motion for the aircraft. The mathematical model of an airplane in the plane x-y is as follows:

$$p' = I_1(-q + l) \tag{8.35}$$
$$q' = I_2(p + m)$$

where I_1 and I_2 are the inertia moments of the airplane with respect to axis x and y, respectively, l and m are physical constants specific to the airplane, and p, q are the positions with respect to axis x and y, respectively. For small velocities, it maybe sufficient to approximate the behavior of an airplane with the model given by Equation (8.35), which ignores the z-component. However, a more realistic mathematical model of an airplane, in three-dimensional space, is as follows:

$$p' = I_1(-qr + l) \qquad (8.36)$$
$$q' = I_2(pr + m)$$
$$r' = I_3(-pq + n)$$

where now I_3 is the inertia moment of the airplane with respect to the z axis, n is a physical constant specific to the airplane, and r is the position along the z axis (see Figure 8.20). We have to mention here that Equation (8.36) consists of a system of three simultaneous non-linear differential equations with very complicated dynamics. This is not the case for Equation (8.35) which is linear and no complicated behavior can occur.

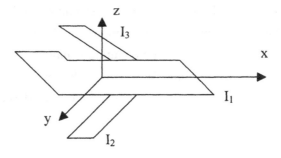

Figure 8.20 An airplane in three-dimensional space.

Next we introduce the influences of the environment. We will be confined to winds for simplicity. Wind disturbances are assumed to have only one component of constant velocity. The magnitude of the constant velocity component is a function of altitude. The constant velocity wind component exists only in the horizontal direction and its value is given in Equation (8.37) as a logarithmic variation with altitude (Jorgensen & Schley, 1995).

$$u_g = u_{wind510} \left[1 + \frac{(\ln(r/510))}{\ln(51)} \right] \qquad (8.37)$$

where: u_g is the constant velocity component, $u_{wind510}$ is the wind speed at 510 ft. altitude (typical value = 20 ft/sec), and r is the aircraft altitude. Then the mathematical model for the airplane with wind disturbances is as follows:

$$p' = I_1(-qr + 1) - u_g \qquad (8.38)$$
$$q' = I_2(pr + m)$$
$$r' = I_3(-pq + n)$$

In Equation (8.38) the wind velocity is affecting only the velocity of the airplane in the x direction. The magnitude of wind velocity is dependent on the altitude r of the airplane. Other disturbances like temperature, pressure and turbulence can also be modeled and could be introduced in the mathematical model of Equation (8.36).

The method for automated mathematical modelling of dynamical systems using the fuzzy-fractal approach (of Chapter 5) can be used to automate the process of modelling aircraft systems. We can use the fractal dimension of the time series of the positions p,q,r for the airplane as a measure of the complexity of the modelling problem. Then, a set of fuzzy rules for model selection has to be developed to decide which models are the most appropriate ones for the airplane. Finally, the best mathematical model for the airplane has to be selected. The interested reader can follow the same procedure used in previous sections (for robotic systems and biochemical reactors) to develop an intelligent system for automated modelling of aircraft dynamic systems.

8.4.2 Simulation results of aircraft systems

We show in this section some simulation results for aircraft systems obtained using the method for automated simulation of dynamical systems (described in Chapter 6). The fuzzy-genetic approach for simulation enables the automated generation of parameter values for the models and obtains the corresponding identifications of the dynamic behaviors. We will show here only simulation results for the case of an airplane in three-dimensional space (Equation (8.36)) and leave to the reader the simulations of other cases.

In Figures 8.21 and 8.22 we show the simulation results for an airplane in three-dimensional space with inertia moments:

$$I_1 = 0.9, \qquad I_2 = 0.5, \qquad I_3 = 0.1$$

and the physical constants are: $l = m = n = 0.1$. The initial conditions are: $p(0) = 0$, $q(0) = 0$, $r(0) = 0$. In Figure 8.21 we show the position p of the airplane plotted from time 0 to 200. We can see clearly, in this figure, how the dynamic behavior of the system is becoming chaotic by period doublings (bifurcations). In Figure 8.22 we show the position q of the airplane plotted again from 0 to 200. We can see from this figure a similar behavior for variable q, but even more chaotic because there more unstable points.

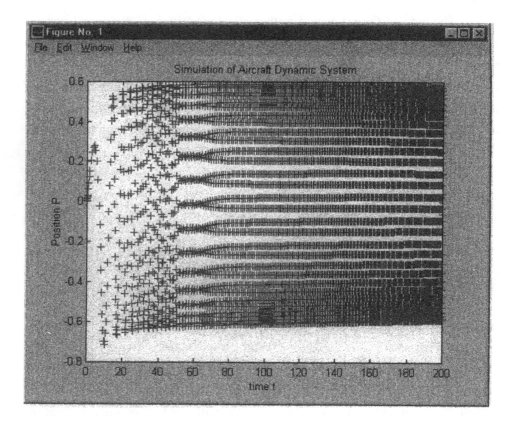

Figure 8.21 Simulation of position p for an airplane with
$I_1 = 0.9, I_2 = 0.5, I_3 = 0.1$

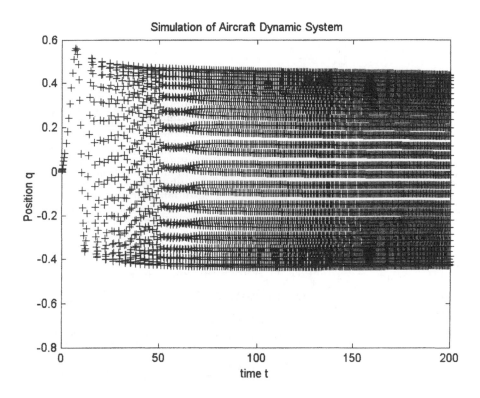

Figure 8.22 Simulation of position q for an airplane with
$$I_1 = 0.9, I_2 = 0.5, I_3 = 0.1$$

In Figure 8.23 we show the simulation results for a smaller airplane with inertia moments:

$$I_1 = 0.3, \qquad I_2 = 0.2, \qquad I_3 = 0.1$$

with the same physical constants and initial conditions. In Figure 8.23 we show the position p of the airplane plotted from time 0 to 200. We can see clearly, in this figure, how the dynamic behavior of the system is becoming chaotic by period doublings but at a slower rate than in the first case. This is because a smaller airplane is more stable than a bigger airplane.

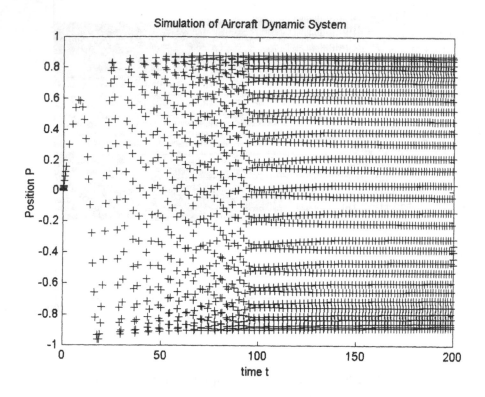

Figure 8.23 Simulation of position p for an airplane with
$$I_1 = 0.3, I_2 = 0.2, I_3 = 0.1$$

In Figure 8.24 we show the simulation results for a large airplane with the following inertia moments:

$$I_1 = 5, \qquad I_2 = 4, \qquad I_3 = 3$$

with the same physical constants and initial conditions. In Figure 8.24 we show the position p of the airplane plotted from time 0 to 100. We can see very clearly, in this figure, how the dynamic behavior of the system is becoming chaotic at a faster rate than in the other cases. This is because we are considering a bigger airplane in this case.

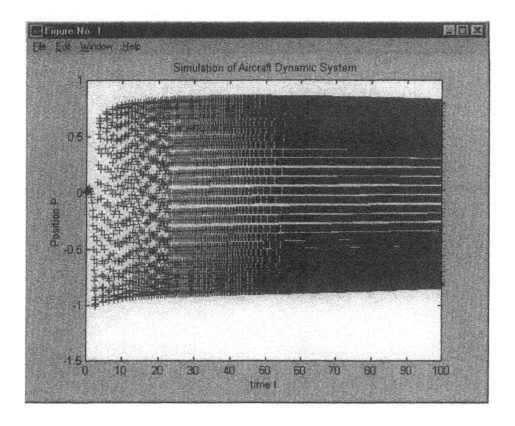

Figure 8.24 Simulation of position p for an airplane with $I_1 = 5, I_2 = 4, I_3 = 3$

In Figure 8.25 we show the simulation results for a smaller airplane with inertia moments:

$$I_1 = 0.009, \qquad I_2 = 0.005, \qquad I_3 = 0.001$$

with the same physical constants and initial conditions. In Figure 8.25 we show the position p of the airplane plotted from time 0 to 1000. We can see in this case how the dynamic behavior is stable for most of this time interval. The reason is that the airplane is so small that instability is highly improbable.

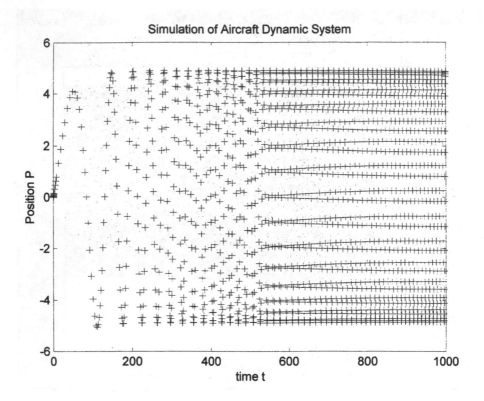

Figure 8.25 Simulation of position p for an airplane with
$I_1 = 0.009$, $I_2 = 0.005$, $I_3 = 0.001$

Now we will consider changing the physical constants l, m, and n of the model of an airplane. In Figure 8.26 we show the simulation results for $I_1 = 0.9$, $I_2 = 0.5$ and $I_3 = 0.1$ and we consider increasing the values of l, m and n to explore the change in dynamic behavior (Figure 8.26).

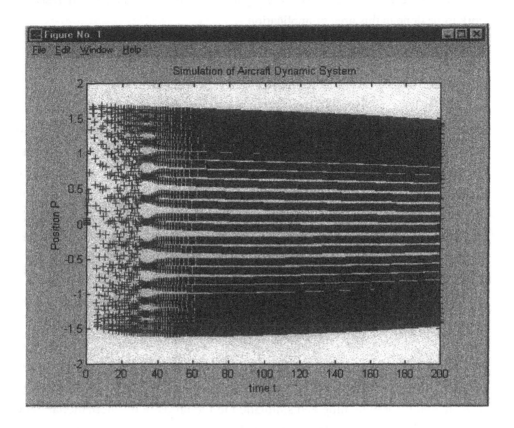

Figure 8.26 Simulation of position p for an airplane with l = 1, m = 1, n = 1

We can see in Figure 8.26, how the dynamic behavior of the system becomes chaotic even faster than in previous cases. The reason for this fact is that the system is more unstable for larger values of the physical constants l, m and n. If we, on the other hand, decrease the values of the constants l, m and n, we can see that dynamic behavior become stable.

Finally, we have to mention that chaotic behavior for the case of aircraft systems has been associated with the dangerous "flutter" effect that occurs in real-world airplanes (Melin & Castillo, 1998c). For this reason, it is very important to understand how and when chaotic behavior occurs for this type of dynamical systems. However, we also have to recognize that still there is a lot of research work to be done in this area of application.

8.5 Concluding Remarks and Future Directions

In this chapter, we have presented several advanced applications of the methods for automated modelling and simulation (described in Chapters 5 and 6) with very good results. First, we described the application of the methods for modelling and simulation to robotic dynamic systems, which is a very interesting and important domain of application. The results were presented only for single-link and two-link robot arms, however the reader is welcome to explore more complicated systems with the methodology presented here. We also described the application of the methods for modelling and simulation to the problem of understanding the complex dynamic behavior of biochemical reactors. We showed mathematical models for biochemical reactors and also simulation results to explore the dynamic behavior of these dynamic systems. We expect the reader to explore similar systems (chemical reactors or nuclear reactors) with the same methodology and obtain also good results. We also describe briefly the application of the methods for modelling and simulation to the problem of international trade between three or more countries. This application is from the area of Economics and poses some difficult questions about the stability of a system of three or more countries with international trade. Finally, we have also considered briefly the problem of modelling and simulation of aircraft systems. We showed some simulation results for aircraft systems and leave to the reader further exploration of this type of dynamical systems. In conclusion, we have to say that we have presented four interesting applications with some encouraging results in the modelling and simulation of the corresponding dynamical systems, but still a lot of research work remains be done with these applications or with similar ones.

Chapter 9

Advanced Applications of Adaptive Model-Based Control

In this chapter, we present several advanced applications of the new method for adaptive model-based control described in Chapter 7 of this book. First, we describe the application of the new method for adaptive model-based control to the case of robotic dynamic systems, which is very important for solving the problem of controlling real-world manipulators in real-time. Second, we describe the application of the method for adaptive model-based control to the case of biochemical reactors in the food industry, which is a very interesting case due to the complexity of this non-linear problem. Third, we consider briefly the problem of controlling international trade between three or more countries, with our new method for adaptive model-based control. Finally, we also consider briefly the problem of controlling aircrafts with our new method for adaptive model-based control. We conclude this chapter with some concluding remarks and also some future directions of research work.

9.1 Intelligent Control of Robotic Dynamic Systems

Given the dynamic equations of motion of a manipulator, the purpose of robot arm control is to maintain the dynamic response of the manipulator in accordance with some prespecified performance criterion (Fu, Gonzalez & Lee, 1987). Although the control problem can be stated in such a simple manner, its solution is complicated by inertial forces, coupling reaction forces, and gravity loading on the links. In general, the control problem consists of (1) obtaining dynamic models of the robotic system, and (2) using these models to determine control laws or strategies to achieve the desired system response and performance. The first part of the control problem has been discussed to some extent in Section 8.1 of the previous chapter. Now, this section concentrates on the latter part of the control problem.

Among various adaptive control methods, the model-based adaptive control is the most widely used and it is also relatively easy to implement. The concept of model-based adaptive control is based on selecting an appropriate reference model and adaptation algorithm which modifies the feedback gains to the actuators of the actual system. The adaptation algorithm is driven by the errors between the reference model outputs and the actual system outputs. A general control block diagram of the model-based adaptive control system is shown in Figure 9.1

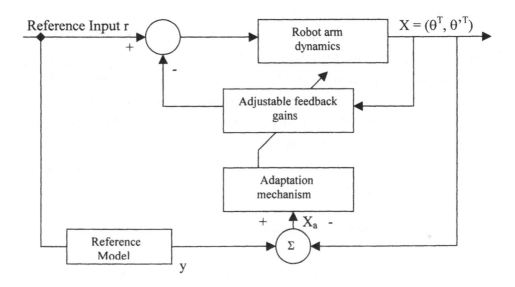

Figure 9.1 A general control block diagram for model-based adaptive control.

9.1.1 Traditional model-based adaptive control of robotic systems

Many authors (Fu, Gonzalez & Lee, 1987) have proposed linear mathematical models to be used as reference models in the general scheme shown in Figure 9.1. For example, a linear second-order time invariant differential equation can be used as the reference model for each degree of freedom of the robot arm. Defining the vector y(t) to represent the reference model response and the vector x(t) to represent the manipulator response, the joint i of the reference model can be described by

$$a_i y''_i(t) + b_i y'_i(t) + y_i(t) = r_i(t) \qquad (9.1)$$

If we assume that the manipulator is controlled by position and velocity feedback gains, and the coupling terms are negligible, then the manipulator dynamic equation for joint i can be written as

$$\alpha_i(t)x''_i(t) + \beta_i(t)x'_i(t) + x_i(t) = r_i(t) \qquad (9.2)$$

where the system parameters $\alpha_i(t)$ and $\beta_i(t)$ are assumed to vary slowly with time.

Several techniques are available to adjust the feedback gains of the controlled system. Due to its simplicity, a steepest descent method is used to minimize a quadratic function of the system error, which is the difference between the response of the actual system (Equation (9.2)) and the response of the reference model (Equation (9.1)).

The fact that this control approach is not dependent on a complex mathematical model is one of its major advantages, but stability considerations of the closed-loop adaptive system are critical. A stability analysis is difficult and has only been carried out using linearized models. However, the adaptability of the controller can become questionable if the interaction forces among the various joints are severe (non-linear).

9.1.2 Adaptive model-based control of robotic systems with a neuro-fuzzy approach

We can apply our new method for adaptive model-based control using a neuro-fuzzy approach (described in Chapter 7) to the problem of controlling robotic dynamic systems. Intelligent control of robotic systems is a difficult problem because the dynamics of these systems is highly non-linear. Optimal control of many robotic systems also requires methods which make use of predictions of future behavior. We describe, in this section, an intelligent control system for controlling robot manipulators to illustrate our neuro-fuzzy hybrid approach for adaptive control.

We use a fuzzy rule base for model selection for the case of robotic manipulators. In Section 8.1, we presented mathematical models that can be used to model the dynamic behavior of robotic manipulators. Lets call M_1 the mathematical model given by Equation (8.13), M_2 the mathematical model given by Equation (8.14), M_3 the model given by Equation (8.15), and M_4 the model given by Equation (8.16). Then we can establish a fuzzy rule base for these models as explained in Section 7.3 of this book. We will assume here without loss of generality that the selection parameters are the fractal dimension of a time series of measured values of the relevant variables in the problem (angle, angular velocity) and the number of links of the manipulator. Also, we are assuming that only four models are needed to model completely the robotic system. Then, we can define a set of four fuzzy if-then rules that basically relate the fuzzy values of the selection parameters with the corresponding mathematical model. We show in Table 9.1 this set of fuzzy rules for model selection for the case of manipulators of one and two links.

Table 9.1 Fuzzy rule base for model selection of robotic systems

IF		THEN
Fractal dimension	Number of links	Mathematical Model
low	one	M_1
high	one	M_2
low	two	M_3
high	two	M_4

We also need to define the membership functions for the fuzzy values in Table 9.1. The membership functions for the models should give us the degree of belief that a particular mathematical model is the correct one for the specific values of the selection parameters. We have to note here that for using a fuzzy rule base (like the one described in Table 9.1) with mathematical models, we need to use our new fuzzy inference system for multiple differential equations (described in Chapter 7).

We use neural networks for identification and control of the robotic dynamic system. The neural networks are trained with the backpropagation algorithm with real data to achieve the desired level of performance. Two multilayer neural networks are used, one for identification of the model of the robotic system and the second for the controller. If we combine the fuzzy rule base for model selection with the neural networks for identification and control, we can obtain an intelligent system for adaptive model-based control of robotic dynamic systems. The intelligent control system combines the advantages of neural networks (ability for identification and control) with the advantages of fuzzy logic (use of expert knowledge) to achieve the goal of robust adaptive control of robotic dynamic systems. The general architecture of the intelligent control system for robotic systems is shown in Figure 9.2. In this figure, we have a module for the fuzzy-rule base of model selection, a module for the neural network of control, and a module for the neural network of identification.

An intelligent control system with the architecture shown in Figure 9.2 is capable of adapting to changing dynamic conditions in the robotic system, because it can change the control actions (given by the network Nc) according to the data measured on-line and also can change the reference mathematical model if there is a large enough change in the fractal dimension of the time series. Of course, a change in the reference mathematical model also causes that the neural network Ni performs a new identification for the model. In conclusion, the intelligent system with the architecture shown in Figure 9.2 achieves model-based control of robotic systems using a combination of Neural Networks and Fuzzy Logic.

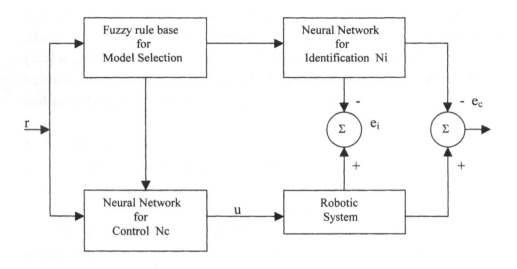

Figure 9.2 General architecture of the intelligent control system for
robotic dynamic systems.

To give an idea of the performance of our neuro-fuzzy approach for
adaptive model-based control of robotic systems, we show below simulation
results obtained for a single-link robot. The desired trajectory for the link was
selected to be

$$q_d = 1.0\sin(2.0(1-e^{-t^3})t)$$

and the simulation was carried out with the initial values:

$$q(0) = 0.1 \qquad q'_1(0) = 0$$

We used three-layer neural networks (with 5 hidden neurons) with the
backpropagation algorithm and hyperbolic tangent sigmoidal functions as the
activation functions for the neurons. We show in Figure 9.3 the initial function
approximation achieved with the neural network for control. Of course, the
approximation is not good (at the beginning) because the net hasn't been trained
yet with the data.

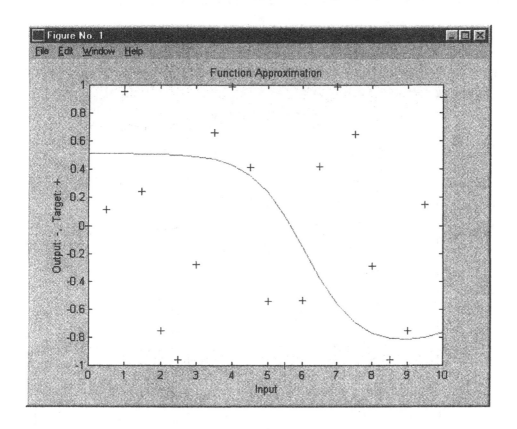

Figure 9.3 Initial function approximation of the neural network for control.

We show in Figure 9.4 the function approximation achieved with the neural network for control after 400 epochs of training with a variable learning rate. The identification achieved by the neural network (after 400 epochs) can be considered very good because the error has been decreased to the order of 10^{-1}. Still, we can obtain a better approximation by using more hidden neurons or more layers. In any case, we can see clearly how the neural network learns to control the robotic system, so it is able to follow the arbitrary desired trajectory.

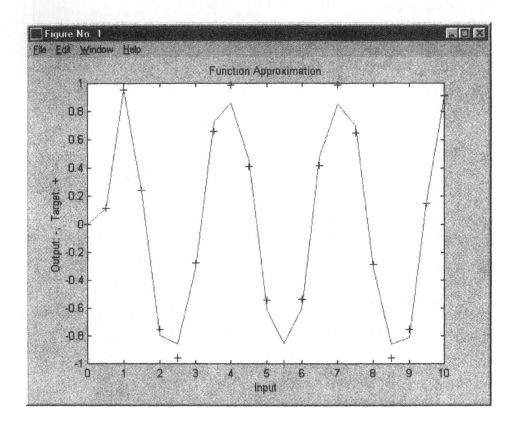

Figure 9.4 Function approximation of the neural network
for control after 400 epochs.

We also show in Figure 9.5 the curve relating the sum of squared errors SSE against the number of epochs of neural network training. We can see in Figure 9.5 how the SSE diminishes rapidly from being of the order of 10^1 to a smaller value of the order of 10^{-1}.

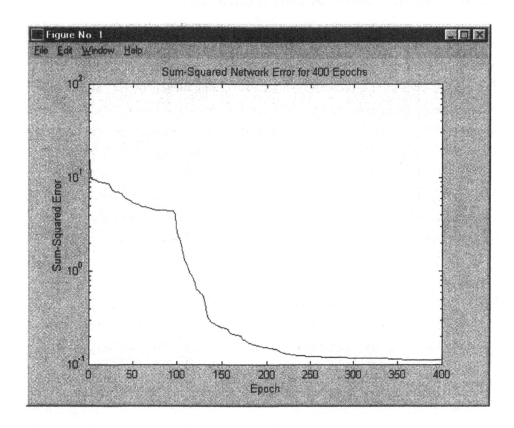

Figure 9.5 Sum of squares of errors for the neural network plotted
from 0 to 400 epochs.

We have to mention here that these simulation experiments for a single
link robot show very good results. We have also tried our approach for control
with more complex robotic systems with encouraging results. We recommend for
the interested reader to follow our methodology for control with this type of
systems, so he or she can get all of the ideas behind solving the problem of robot
control.

9.2 Intelligent Control of Biochemical Reactors

We describe in this section the application of our new method for adaptive model-based control to the complex case of biochemical reactors in the food industry. The case of biochemical reactors for food processing plants is a very complex one because biochemical processes are often highly non-linear and difficult to control (Ungar, 1995). Optimal control of many biochemical processes also requires systems which make use of predictions of future behavior. In this section, we describe the methodology to develop and intelligent control system for biochemical reactors that can be used in food processing plants to maximize food production by controlling the biochemical processes that occur in the biochemical reactors (Melin & Castillo, 1998b).

9.2.1 Fuzzy rule base for model selection

We describe in this section a fuzzy rule base for model selection for the case of biochemical reactors producing yogurt. In Section 8.2, we presented the mathematical models that can be used to model the dynamical behavior in the biochemical reactors for this case. Lets call M_1 the mathematical model given by Equation (8.21), M_2 the mathematical model given by Equation (8.22), and M_3 the mathematical model given by Equation (8.23). Then we can establish a fuzzy rule base for these models as explained in Section 7.3 of this book. We will assume in the following that the selection parameter is the temperature T used in the production process, defined over the real-valued interval:

$$100 \leq T \leq 120$$

because the range of temperatures used in the production process of yogurt is usually between 100 °F and 120 °F. Since we have three mathematical models in this case, we define three subintervals of [100, 120] as follows:

$$100 \leq T < 105, 105 \leq T < 115, 115 \leq T \leq 120,$$

where M_1 corresponds to the first subinterval, M_2 corresponds to the second subinterval, and M_3 corresponds to the third subinterval. Then, we can define a set of three fuzzy if-then rules that basically relate the subintervals to the mathematical models in a one-to-one fashion. We show the set of fuzzy rules for model selection in Table 9.2.

Table 9.2 Fuzzy rule base for model selection.

IF		THEN
$100 \le T < 105$	(Low)	Mathematical_Model = M_1
$105 \le T < 115$	(Medium)	Mathematical_Model = M_2
$115 \le T \le 120$	(High)	Mathematical_Model = M_3

We also need to define the membership functions for the three corresponding mathematical models. The membership functions for the models should give us the degree of belief that a particular mathematical model is the correct one for a specific value of the temperature T in the closed interval [100,120]. In Figure 9.6 we show the membership functions for models M_1, M_2 and M_3.

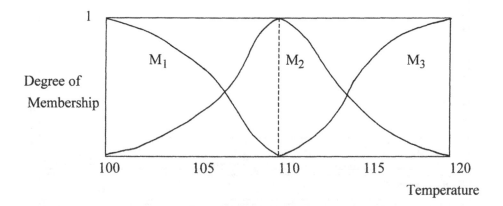

Figure 9.6 Membership functions for the mathematical models.

In Figure 9.6, the membership functions for models M_1 and M_3 are of the "sigmoidal" type and the membership function for model M_2 is of the "gaussian" type. This is to guarantee a smooth transition between the degree of membership between the different mathematical models.

We have implemented this fuzzy rule base in the MATLAB programming language (the complete program is listed in Appendix C). The MATLAB programming language has symbolic and numeric features (Hanselman & Littlefield, 1995). Also the MATLAB has available the "Fuzzy Logic Toolbox" which enables an easy implementation of fuzzy inference systems (Jang & Gulley, 1997). We can use the "Rule Editor" of the Fuzzy Toolbox to construct the rules of the fuzzy inference system. In Figure 9.7, we show the fuzzy rules of Table 9.2 as they are entered in the Rule Editor.

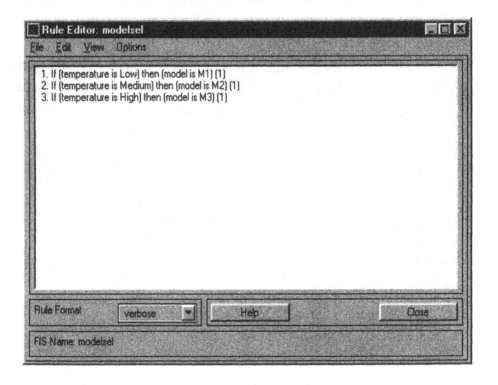

Figure 9.7 Fuzzy rule base for model selection in the Rule Editor.

We can use the "Membership Function Editor" to display and edit all of the membership functions for the fuzzy inference system. In Figure 9.8, we show the membership functions for the mathematical models (output) as they are edited in the Membership Function editor. Also, we show in Figure 9.9 the membership functions for the temperature (input).

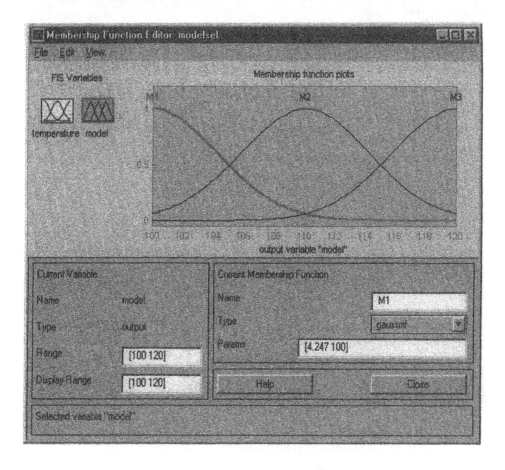

Figure 9.8 Membership functions (model selection) in the
"Membership Function Editor".

In Figure 9.8 we show the three different membership functions for the models M_1, M_2 and M_3 and we can see the "smooth transition" between the degree of membership between each mathematical model.

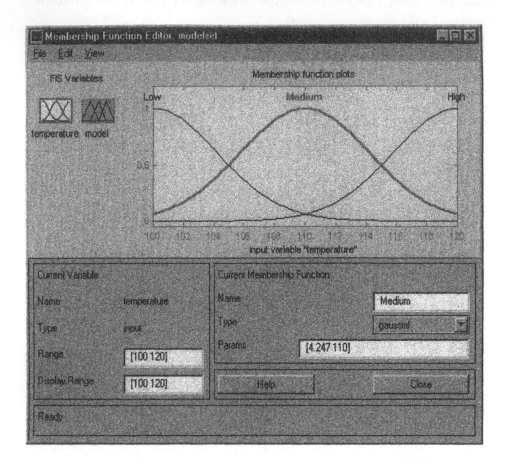

Figure 9.9 Membership functions for the temperature in the
"Membership Function Editor"

In Figure 9.9 we show the three different membership functions for the temperature classified as Low, Medium and High.

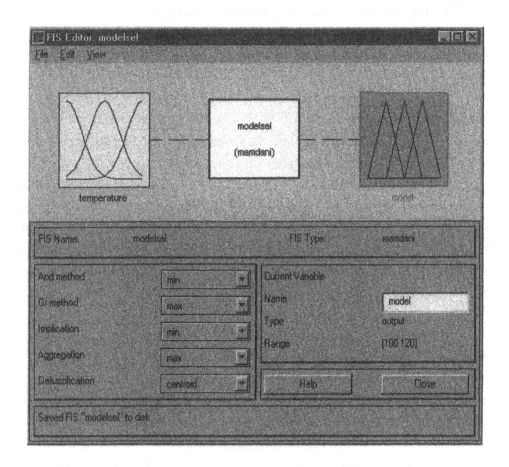

Figure 9.10 General structure of the fuzzy rule base in the FIS Editor.

In the Fuzzy Inference System (FIS) Editor we can display the general information about the fuzzy rule base that we have just designed. In Figure 9.10, we show the general structure of the fuzzy rule base that we have just designed for model selection.

9.2.2 Neural networks for identification and control

We describe in this section the neural networks used for identification and control for the case of adaptive model-based control of biochemical reactors producing yogurt. The neural networks were trained initially using the "backpropagation" learning algorithm (Miller, Sutton & Werbos, 1995) with real data to achieve the desired level of performance and then they were tested for robustness with new data. Process control of biochemical plants is an attractive application because of the potential benefits to both adaptive network research and to actual biochemical process control. In spite of the extensive work on self-tuning controllers and model-reference control, there are many problems in chemical processing industries for which current techniques are inadequate. Many of the limitations of current adaptive controllers arise in trying to control poorly modeled non-linear systems. For most of these processes extensive data are available from past runs, but it has been difficult to formulate precise models (Ungar, 1995).

Bioreactors are difficult to model because of the complexity of the living organisms in them and also they are difficult to control because one often can't measure on-line the concentration of the chemicals being metabolized or produced. Bioreactors can also have markedly different operating regimes, depending on whether the bacteria is rapidly growing or producing product. Model-based control of this reactors offers a dual problem: determining a realistic process model and determining effective control laws in the face of inaccurate process models and highly non-linear processes.

Biochemical systems can be relatively simple in that they have few variables, but still very difficult to control due to strong nonlinearities which are difficult to model accurately. A prime example is the bioreactor. In its simplest form, a bioreactor is simply a tank containing water and cells (e.g. bacteria) which consume nutrients ("substrate") and produce products (both desired and undesired) and more cells. Bioreactors can be quite complex: cells are self-regulatory mechanisms, and adjust their growth rates and production of different products radically depending on temperature and concentrations of waste products. Mathematical models for these systems can be expressed as differential equations of the type shown in Section 8.2 of the previous chapter.

We have implemented a model-based neural controller using the architecture of Figure 7.6, described in Chapter 7. Two multilayer neural networks are used, one for identification of the model of the plant and the second for the controller. Each neural network has 3 layers, a one node input layer, a 5-node hidden layer, and a one-node output layer. We show in Figure 9.11 the architecture of the neural networks for identification and control. The neural networks were implemented in the MATLAB programming language to achieve a high level of efficiency on the numerical calculations needed for these networks. We trained the adaptive networks using temperatures and control actions varying over the range of values relevant to the specific application. The "backpropagation" learning algorithm was used with the data to obtain the weights of the networks.

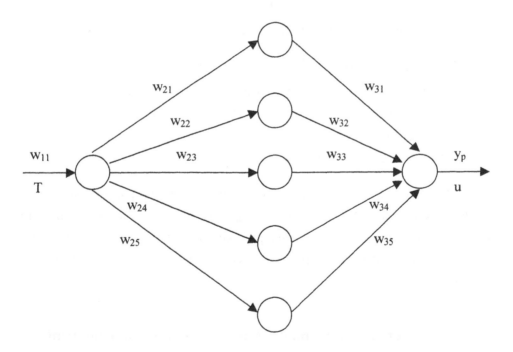

Figure 9.11 Structure of the networks for identification and control.

The backpropagation algorithm (Soucek, 1991) can be written as follows:

1.- Start with a random set of weights.

2.- Calculate y_p by propagating the input x_p trough the network.

3.- Calculate the error of a node corresponding to the pattern p as:

$$Ep = 1/2 \, (d_p - y_p)^2$$

where d_p is the desired output.

4.- Adjust the weights of the network with the iterative equation:

$$W_{ln}(t+1) = W_{ln}(t) + \mu \delta_{pln} \, X_{pl}$$

where δ_{pln} is the error for node n at layer l and for the pattern p, given by the equation:

$$\delta_{pLn} = d_{pLn} - y_{pLn}$$

if node n is an output node, and by equation:

$$\delta_{pln} = f'(y_{pln}) \, \Sigma_r \, \delta_{p,l+1,r} \, W_{l+1,r,n}$$

where r is over the nodes in layer l+1. f(y) is the activation function of the nodes. The activation function can be a sigmoidal function, for example, the logistic function is widely used.

5.- Repeat by going to step 2.

The complete computer program for the backpropagation algorithm, implemented in the MATLAB programming language, can be found in Appendix C of this manuscript. This computer program can be used to train the neural networks with real data for the problem of controlling biochemical reactors in the Food Industry.

9.2.3 Intelligent adaptive model-based control for biochemical reactors

In this section, we combine the implementation of the fuzzy rule base for model selection with the implementation of the neural networks for identification and control to obtain an intelligent system for adaptive model-based control for biochemical reactors. This intelligent control system combines the advantages of neural networks (ability for identification and control) with the advantages of fuzzy logic (use of expert knowledge) to achieve the goal of robust adaptive control of biochemical reactors in food processing plants. The general architecture of the intelligent control system for biochemical reactors is shown in Figure 9.12. In this figure, we have a module for the fuzzy rule base of model selection, a module for the Neural Network of Control, and a module for the Neural network of Identification.

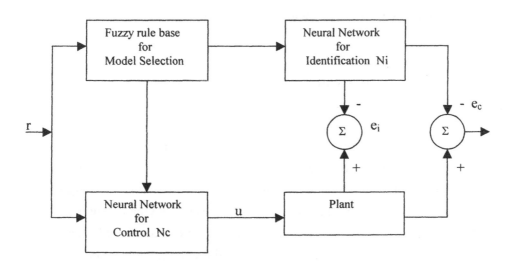

Figure 9.12 General architecture of the intelligent control system

An intelligent control system with the architecture shown in Figure 9.12 is capable of adapting to changing dynamic conditions in the biochemical reactors, because it can change the control actions (given by the network N_c) according to the data measured on-line and also can change the reference mathematical model if there is a large enough change in the temperature T. Of course, a change in the reference mathematical model also causes that the neural network N_i performs a new identification for the model.

After the neural networks were trained, we validated their performance with simulations to get an idea of the degree of approximation achieved. The results can be considered very good because the errors (of identification and control) achieved with the networks were relatively small. In the following figures, we show some of these results to give an idea of the performance of the neural networks for identification and control.

We show in Figure 9.13 the initial function approximation achieved with a neural network for identification with the architecture shown in Figure 9.11. Of course, the approximation is not good (at the beginning) because the net hasn't been trained yet with the data. We set the parameters for the backpropagation algorithm as follows:

$$\text{goal\_error} = 0.00002$$
$$\text{learning\_rate} = 0.0001$$

and we use as activation functions hyperbolic tangent sigmoidal functions (tansig). We use as reference model for the identification, the model of two bacteria (M_2) used for production, because this is the case that is been considered for food production. This is sufficient for our purpose of having a neural network that knows the process of the plant, because the M_1 model (of one bacteria) can be considered a special case of the M_2 model and the M_3 model can be treated as a case to be controlled.

We show in Figure 9.14 the function approximation achieved with the neural network for identification after 40,000 epochs of training with a learning rate of 0.0001. The target values shown in Figure 9.14 are from the numerical solution of the M_2 model (system of coupled differential equations) given as Equation (8.22) in Chapter 8. The curve shown as output in Figure 9.14 is the approximation given by the neural network after training it with the target values of the M2 model.

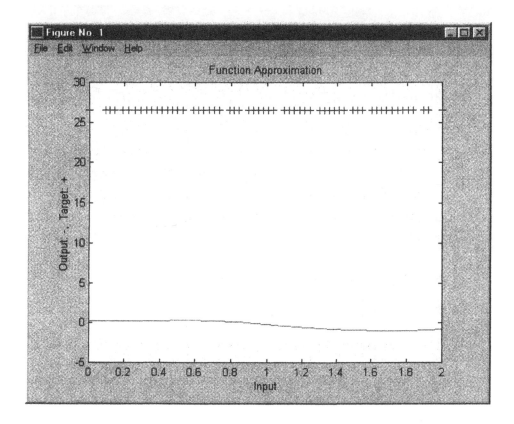

Figure 9.13 Initial function approximation of the neural network for identification.

The identification of the model achieved (after 40,000 epochs) by the neural network, shown in Figure 9.14, can't be considered good at all. This means that still more training is needed to achieve the desired level of accuracy.

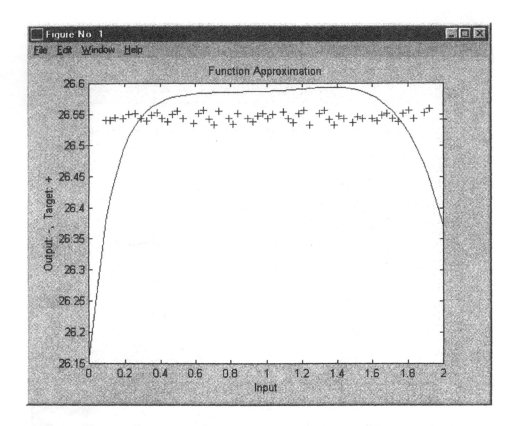

Figure 9.14 Function approximation of the neural network for identification
after 40,000 epochs.

We show in Figure 9.15 the curve relating the sum of Squared Errors
(SSE) against the number of epochs of neural network training. We can see in
Figure 9.15 how the SSE diminishes rapidly from being of the order of 10^5 to the
smaller value of the order of 10^1. However, we need more training to achieve an
even smaller value of SSE.

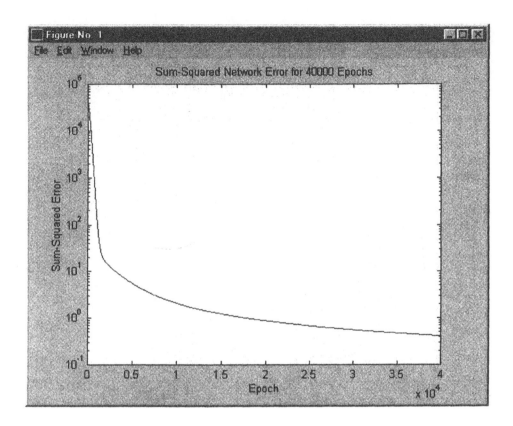

Figure 9.15 Sum of squares of errors for the neural network plotted
from 0 to 40,000 epochs

We show in Figure 9.16 the function approximation achieved with the
network for identification after 80,000 epochs of training with a learning rate of
0.0001. The target values shown in Figure 9.16 are from the numerical solution of
the M_2 model (Equation (8.22)). The curve shown as output in Figure 9.16 is the
approximation of the neural network after training it with the target values. The
identification achieved (after 80,000 epochs) by the neural network is much better
than the one shown in Figure 9.14. Of course, more training could improve even
more the approximation.

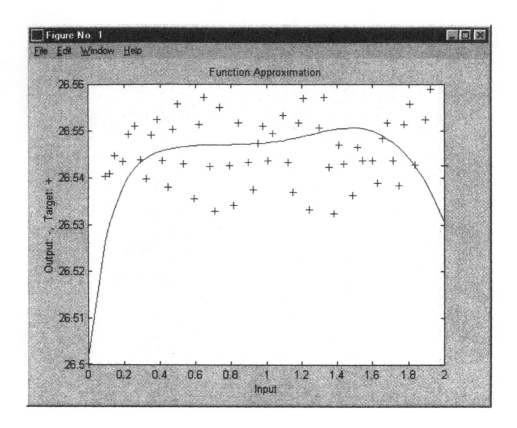

Figure 9.16 Function approximation of the neural network for identification
after 80,000 epochs

We show in Figure 9.17 the curve relating the sum of squared errors against the number of epochs of neural network training. We can see in Figure 9.17 how the sum of squared errors diminishes rapidly from being of the order of 10^5 to the relatively small value of 10^{-2}. The fact that the sum of errors is of the order of 10^{-2} after 80,000 epochs, means that the neural network has achieved a relative good approximation to the solution of the mathematical model M_2. Of course, more training of the neural network could improve even more this approximation. However, we have considered this approximation as sufficiently good for identification of the complex model of two bacteria for food production.

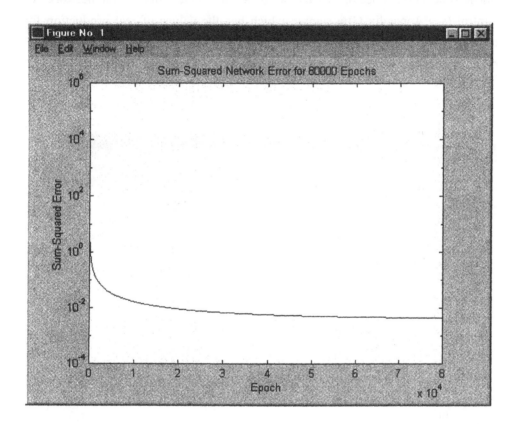

Figure 9.17 Sum of squares of errors for the neural network plotted
from 0 to 80,000 epochs.

We show in Figure 9.18 the initial function approximation achieved with a neural network for control with the architecture shown in Figure 9.11. Of course, the approximation is not good (at the beginning) because the network hasn't been trained yet with the data. The parameters for the backpropagation algorithm are the same as the ones used for identification. On the other hand, we use in this case as activation functions the pure linear ones (purelin), instead of sigmoidal type functions. This is because production control only requires the simulation of a linear production process. We use as data for training the neural network for control (see Appendix C) a sample which has the values of real production for different times. This is sufficient, in this case, for our purpose of having a neural network that knows how to control the production process of the plant.

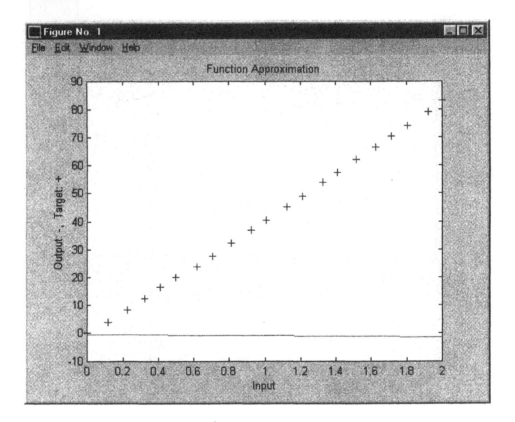

Figure 9.18 Initial function approximation of the neural network for control.

We show in Figure 9.19 the function approximation achieved by the neural network for control after 5,000 epochs of training with a learning rate of 0.0001. The target values shown in Figure 9.19 are from the sample of production data (Appendix C) used to train the neural network. The line shown as output in Figure 9.19 is the approximation of the neural network to the pattern of the target values. The approximation achieved (after 5,000 epochs) by the neural network for control is excellent (as can be seen in Figure 9.19). The sum of squared errors is of the order of 10^{-2}, which is sufficient in this case.

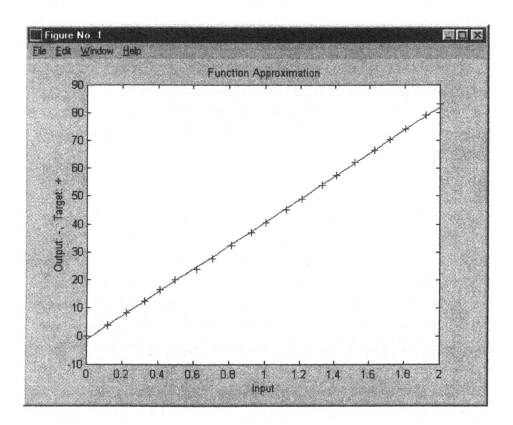

Figure 9.19 Function approximation of the neural network for control
after 5,000 epochs.

We have to remember that this training of the neural networks is only to have an initial knowledge of the plant. Later when the neural networks are used on-line, the real data that is going to be measured will be used to improve even more their performance.

9.3 Intelligent Control of International Trade

We describe in this section the application of our new method for adaptive model-based control to the problem of controlling international trade dynamics. The problem of international trade between three or more countries is a very complex one because of the couplings and non-linearities involved in the mathematical models (Castillo & Melin, 1998c). In this section, we describe the methodology to develop an intelligent system for controlling international trade that can be used by the government of a specific country to maximize the profit from its international trade with other countries.

9.3.1 Adaptive model-based control of international trade

The method for adaptive model-based control of non-linear dynamical systems consists of using a fuzzy rule base for model selection, a neural network for identification and a neural network for control (as described in Chapter 7). For the case of international trade, we need to define each of the method's components mentioned above to achieve the goal of controlling the dynamical system of three (or more) countries with trade between them.

The mathematical models of international trade can be represented as systems of coupled non-linear differential equations (as described in Section 8.3 of the previous chapter). In this case, we can establish a fuzzy rule base for model selection that enables the use of the appropriate mathematical model according to the changing conditions of the economies involved. For example, if we use the general mathematical models of Equations (8.28) and (8.29) for describing the international trade dynamics between one, two or three countries, we can have the following specific models. For one country with no international trade we have:

M_1: $\quad\quad y'_1 = \alpha_1(I_1 - S_1)$

$\quad\quad\quad\quad r'_1 = \beta_1(L_1 - M_1/p_1)$

For two countries with no international trade:

M_2: $\quad\quad y'_i = \alpha_i(I_i - S_i)$ $\quad\quad\quad\quad\quad\quad\quad i = 1,2$

$\quad\quad\quad\quad r'_i = \beta_i(L_i - M_i/p_i)$

For two countries with international trade:

M_3: $\quad\quad y'_i = \alpha_i(I_i - S_i + \gamma\,(EX_i - IM_i\,))$ $\quad\quad i = 1,2$

$\quad\quad\quad\quad r'_i = \beta_i(L_i - M_i/p_i)$

For three countries with no international trade:

M_4: $\quad\quad y'_i = \alpha_i(I_i - S_i)$ $\quad\quad\quad\quad\quad\quad\quad i = 1,2,3$

$\quad\quad\quad\quad r'_i = \beta_i(L_i - M_i/p_i)$

And for three countries with international trade:

M_5: $\quad\quad y'_i = \alpha_i(I_i - S_i + \gamma\,(EX_i - IM_i\,))$ $\quad\quad i = 1,2,3$

$\quad\quad\quad\quad r'_i = \beta_i(L_i - M_i/p_i)$

where I_i, S_i, L_i, M_i, EX_i, IM_i, and p_i are defined as in Section 8.3 of the previous chapter. Now, using γ as a selection parameter we can establish the fuzzy rule base for model selection as in Table 9.3.

Table 9.3 Fuzzy rule base for model selection of international trade

IF		THEN
γ	Number of countries	Mathematical Model
	one	M_1
small	two	M_2
large	two	M_3
small	three	M_4
large	three	M_5

In Table 9.3 we are assuming that the selection parameter γ can have only two possible fuzzy values (small and large). The reasoning behind this is that when γ is small, we can use the model with no international trade and when γ is large we can use the model with international trade. We have to note here that the fuzzy rule base has to be developed according to the particular case that is being considered.

We use neural networks, for identification and control, trained with the backpropagation algorithm (as in previous section). The integration of the fuzzy rule base for model selection with the neural networks for identification and control, results in an intelligent system for adaptive model-based control of international trade. This intelligent system combines the advantages of neural networks (ability for identification and control) with the advantages of fuzzy logic (use of expert knowledge) to achieve the goal of robust adaptive control of international trade. The general architecture of the intelligent control system for international trade is similar to the one shown in Figure 9.12, except that now instead of the plant we have a non-linear dynamical system in economics. An intelligent system with this architecture is capable of adapting to changing conditions in the economies of the countries, because it can change the control actions according to the data available and also can change the reference mathematical model if there is a large enough change in the parameter γ. Of course, for this method to work we need to estimate parameter γ from time series of the real values for the variables in the mathematical models.

9.3.2 Simulation results for control of international trade

To give an idea of the performance of our neuro-fuzzy approach for adaptive model-based control of international trade dynamics, we show below simulation results obtained for the case of three countries (USA, Canada and Mexico) with international trade. We will consider the problem of controlling the economy of the less developed country (Mexico) because it is the most challenging from the control point of view. For the case of Mexico, one problem is that of reducing interest rates in the short term so we will consider as a desired trajectory for this economy:

$$r_d = 0.25e^{-0.1t} + 0.02\sin t + 0.05$$

with initial values of :

$$r(0) = 0.30 \qquad r'(0) = 0.$$

In this desired trajectory for the economy, we are assuming that the goal interest rate is 5% and that we need to decrease the initial rate of 30% to the final interest rate of 5%. We also consider that the economy has natural cycles and because of this fact we use the "sine' function.

We use three-layer neural networks (with 10 hidden neurons) with the backpropagation algorithm and hyperbolic tangent sigmoidal functions as the activation functions for the neurons. We show in Figure 9.20 the initial function approximation achieved with the neural networks for control.

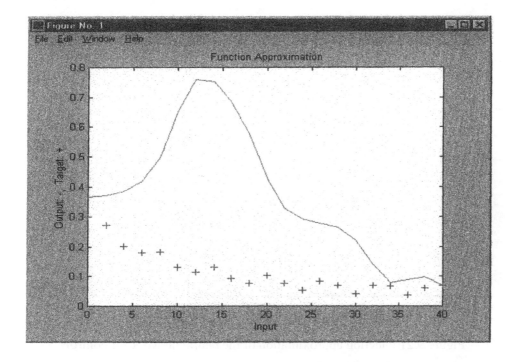

Figure 9.20 Initial function approximation of the neural network for control.

We show in Figure 9.21 the function approximation achieved with the neural network for control after 59 epochs of training with a variable learning rate. The identification achieved by the neural network (after 59 epochs) can be considered very good because the error has been decreased to the order of 10^{-4}. Still, we can obtain a better approximation by using more hidden neurons or more layers. In any case, we can see clearly how the neural network learns to control the economic dynamic system, because it is able to follow the arbitrary desired trajectory.

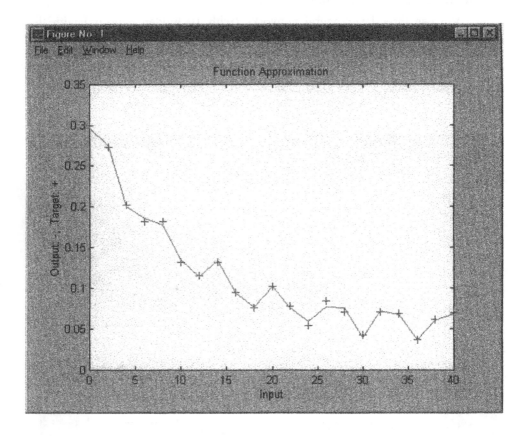

Figure 9.21 Function approximation of the neural network for control after 59 epochs.

We also show in Figure 9.22 the curve relating the sum of squared errors SSE against the number of epochs of neural network training. We can see in Figure 9.22 how the SSE decreases rapidly from being of the order of 10^1 to a smaller value of the order of 10^{-4}.

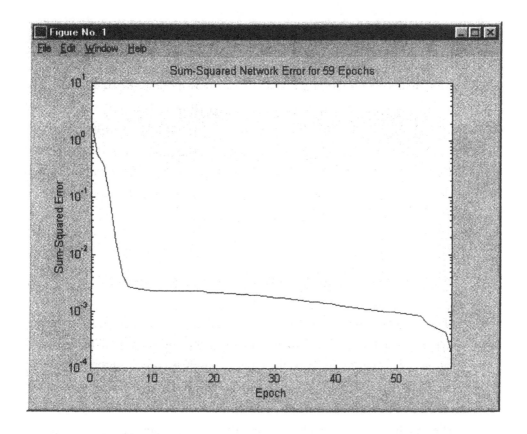

Figure 9.22 Sum of squares of errors for the neural network plotted
from 0 to 59 epochs.

We have to mention here that these simulation experiments for the case of three specific countries (USA, Canada and Mexico) show very good results. We have also tried our approach for control with other dynamic systems in economics with encouraging results. We recommend to the interested reader to use our methodology for this type of economic systems or other similar systems to explore on his (or her) own the interesting problem of controlling complex non-linear dynamical systems.

9.4 Intelligent Control of Aircraft Dynamic Systems

We describe in this section the application of our new method for adaptive model-based control to the problem of controlling the dynamics of aircraft systems. The problem of controlling the dynamic behavior of an aircraft during flight is a very complex one because of the strong non-linearities involved in the mathematical models (Melin & Castillo 1998c). In this section, we describe the methodology to develop an intelligent system for controlling aircraft systems that can be used to automate the flight (or part of it) of a real airplane.

9.4.1 Adaptive model-based control of aircraft systems

The method for adaptive model-based control of non-linear dynamical systems consists of using a fuzzy rule base for model selection, a neural network for identification and a neural network for control. For the case of aircraft systems, we need to define each of these components to achieve the goal of controlling the dynamical system during flight.

The mathematical models of aircraft systems can be represented as coupled non-linear differential equations (as described in Section 8.4). In this case, we can develop a fuzzy rule base for model selection that enables the use of the appropriate mathematical model according to the changing conditions of the aircraft and its environment. For example, if we use the general mathematical models of Equations (8.35), (8.36) and (8.38) for describing aircraft dynamics, we can formulate a set of fuzzy if-then rules that relate the models to the conditions of the aircraft and its environment. Lets assume that M_1 is given by Equation (8.35), M_2 is given by Equation (8.36) and M_3 is given by Equation (8.38). Now, using the wind velocity u_g and inertia moment I_1 as selection parameters we can establish the fuzzy rule base for model selection as in Table 9.4.

Table 9.4 Fuzzy rule base for model selection of aircraft systems

IF		THEN
Wind Velocity, u_g	Inertia Moment, I_1	Mathematical Model
small	small	M_1
small	large	M_2
large	large	M_3

In Table 9.4, we are assuming that the wind velocity u_g can have only two possible fuzzy values (small and large). This is sufficient to know if we have to use the mathematical model that takes into account the effect of the wind (M_3) for u_g large, or if we don't need to use it and simply the model M_2 is sufficient (for u_g small). Also, the inertia moment (I_1) helps in deciding between models M_1 and M_2 (or M_3).

We use neural networks, for identification and control, trained with the backpropagation algorithm (as in previous sections). The integration of the fuzzy rule base for model selection with the neural networks for identification and control, results in an intelligent system for adaptive model-based control of aircraft dynamic systems. This intelligent system combines the advantages of neural networks (ability for identification and control) with the advantages of fuzzy logic (use of expert knowledge) to achieve the goal of robust adaptive control of aircrafts. The general architecture of the intelligent control system for aircrafts is similar to the one shown in Figure 9.12, except that now instead of the plant we have an aircraft dynamic system. An intelligent system with this architecture is capable of adapting to changing conditions in the airplane or in its environment, because it can change the control actions according to the data available and also can change the reference mathematical model if there is a large enough change in the parameters u_g and I_1. Of course, for this method to work we need to estimate these parameters from the time series of the real values for the variables in the mathematical models.

9.4.2 Simulation results for control of aircraft systems

To give an idea of the performance of our neuro-fuzzy approach for adaptive model-based control of aircraft dynamics, we show below simulation results obtained for the case of controlling the altitude of an airplane for a flight of 5 hours. We assume that the airplane takes about one hour to achieve the cruising altitude 30 000 ft, then cruises along for about three hours at this altitude (with minor fluctuations), and finally descends for about one hour to its final landing point. We will consider the desired trajectory as follows:

$$r_d = \begin{cases} 30t & \text{for } 0 \leq t \leq 1 \\ 30 + 2\sin 10t & \text{for } 1 < t \leq 4 \\ 150 - 30t & \text{for } 4 < t \leq 5 \end{cases}$$

Of course, a complete desired trajectory for the airplane would have to include the positions for the airplane in the x and y directions (variables p, q in the models of Section 8.4). However, we think that here for illustration purposes is sufficient to show the control of the altitude r for the airplane.

We used three-layer neural networks (with 10 hidden neurons) with the backpropagation algorithm and hyperbolic tangent sigmoidal functions as the activation functions for the neurons. We show in Figure 9.23 the initial function approximation achieved with the neural network for control.

We show in Figure 9.24 the function approximation achieved by the neural network for control after 600 epochs of training with a variable learning rate. The identification achieved by the neural network (after 600 epochs) can be considered very good because the error has been decreased to the order of 10^1. Still, we can obtain a better approximation by using more hidden neurons or more layers. In any case, we can see clearly (from Figure 9.24) how the neural network learns to control the aircraft, because it is able to follow the arbitrary desired trajectory.

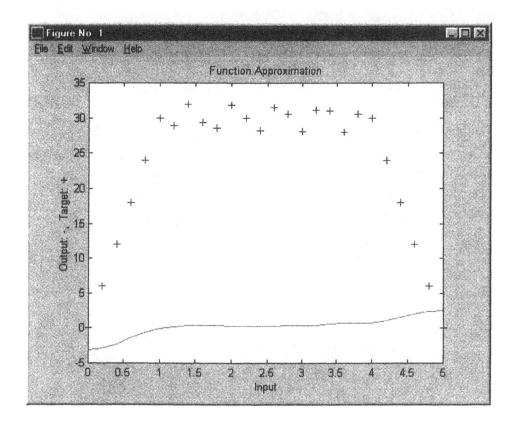

Figure 9.23 Initial function approximation of the neural network
for control of an airplane.

We also show in Figure 9.25 the curve relating the sum of squared errors
SSE against the number of epochs of neural network training. We can see in
Figure 9.25 how the SSE decreases rapidly from being of the order of 10^4 to a
smaller value of the order of 10^1.

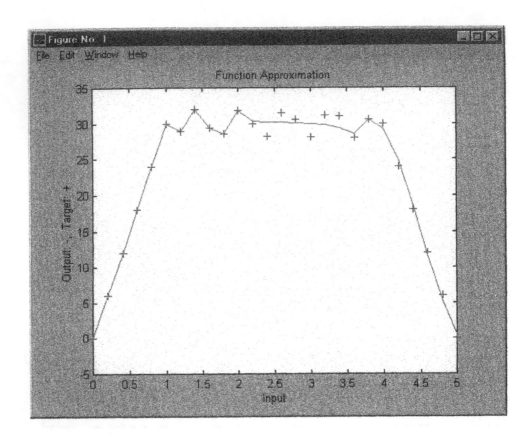

Figure 9.24 Function approximation of the neural network for control
of an airplane after 600 epochs.

We have to mention here that these simulation experiments for the case of a specific flight for a given airplane show very good results. We have also tried our approach for control with other types of flights and airplanes with encouraging results. We leave to the reader further experimentation with this type of aircraft systems and other similar dynamical systems.

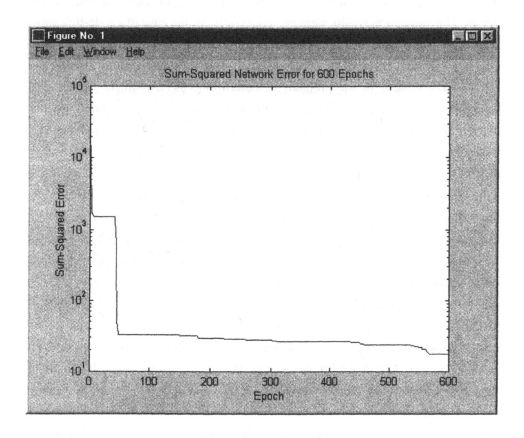

Figure 9.25 Sum of squares of errors for the neural network plotted
from 0 to 600 epochs.

9.5 Concluding Remarks and Future Directions

In this chapter, we have presented several advanced applications of the method for adaptive model-based control (described in Chapter 7) with very good results. First, we described the application of the method for adaptive model-based control to the problem of controlling robotic dynamic systems, which is a very important domain of application for areas such as manufacturing, medicine, aerospace and others. The simulation results were presented only for the case of single-link robot arms, however the reader is welcome to explore more complicated systems with the methodology presented here. We also described the application of the method for adaptive control to the problem of controlling the dynamical behavior of biochemical reactors used for food production in the food industry. The simulation results we presented only for relative simple biochemical reactors (for the case of producing yogurt), however we expect the reader to explore the control of similar systems (like chemical reactors or nuclear reactors) with the same methodology and obtain also good results. We also described briefly the application of the method for adaptive control to the problem of controlling international trade between several countries. This application is from the area of Economics and poses some difficult questions about the stabilization (or control) of an erratic economy with international trade. We have encouraging results in this area of application, but still there is a lot of work to be done for this type of problems. Finally, we have also considered briefly the problem of controlling aircrafts systems during flight. We have showed some simulation results for aircraft systems and leave to the reader further exploration of this type of dynamical systems. In conclusion, we have to say that we have presented four interesting applications of the method for adaptive control with encouraging results in controlling the corresponding dynamical systems, but still a lot of research work remains to be done with these applications or with similar ones.

References

Abraham, E. & Firth, W. J. (1984). "Multiparameter Universal Route to Chaos in a Fabry-Perot Resonator", Optical Bistability, Vol. 2, pp. 119-126.

Albertos, P., Strietzel, R. & Mart, N. (1997). "Control Engineering Solutions: A Practical Approach", IEEE Computer Society Press.

Badiru, A.B. (1992). "Expert Systems Applications in Engineering and Manufacturing", Prentice-Hall.

Barto, A. G., Sutton, R. S. & Anderson, C. (1983). "Neuronlike Elements that can Solve Difficult Learning Control Problems", IEEE Transactions on Systems, Man & Cybernetics, Vol. 13, pp. 835-846.

Bernard, J. A. (1988). "Use of Rule-Based System for Process Control", IEEE Control Systems Magazine, Vol. 8, pp. 3-13.

Bratko, I. (1990). "Prolog Programming for Artificial Intelligence", Addison Wesley.

Brogan, W. (1991). "Modern Control Theory", Prentice-Hall.

Bryson, A. E. & Ho Y.-C. (1969). "Applied Optimal Control", Blaisdell Press.

Castillo, O. & Melin, P. (1994a). "Developing a New Method for the Identification of Microorganisms for the Food Industry using the Fractal Dimension, Journal of Fractals, World Scientific, Vol. 2, No. 3, pp. 457-460.

Castillo, O. & Melin, P. (1994b). "An Intelligent System for Discovering Mathematical Models for Financial Time Series Prediction", Proceedings of TENCON'94, IEEE Computer Society Press, Vol. 1, pp. 217-221.

Castillo, O. & Melin, P. (1995a). "An Intelligent System for Financial Time Series Prediction Combining Dynamical Systems Theory, Fractal Theory and Statistical Methods", Proceedings of CIFER'95, IEEE Computer Society Press, pp.151-155.

Castillo, O. & Melin, P. (1995b). "Intelligent Model Discovery for Financial Time Series Prediction using Non-Linear Dynamical Systems and Statistical Methods", Proceedings of the Third International Conference on Artificial Intelligence Applications on Wall Street", Software Engineering Press, pp. 80-89.

Castillo, O. & Melin, P. (1995c). "An Intelligent System for the Simulation of Non-Linear Dynamical Economical Systems", Journal of Mathematical Modelling and Simulation in Systems Analysis, Edited by Achim Sydow, Gordon and Breach Publishers, Vol. 18-19, pp. 767-770.

Castillo, O. & Melin, P. (1996a). "Automated Mathematical Modelling for Financial Time Series Prediction using Fuzzy Logic, Dynamical Systems and Fractal Theory", Proceedings of CIFER'96, IEEE Computer Society Press, pp. 120-126.

Castillo, O. & Melin, P. (1996b). "Automated Mathematical Modelling and Simulation of Dynamical Engineering Systems using Artificial Intelligence Techniques", Proceedings CESA'96, Gerf EC Lille, pp. 682-687.

Castillo, O. & Melin, P. (1996c). "An Intelligent System for Financial Time Series Prediction using Fuzzy Logic Techniques and Fractal Theory", Proceedings ITHURS'96, Vol. 1, AMSE Press, pp. 423-430.

Castillo, O. & Melin, P. (1997a). "Mathematical Modelling and Simulation of Robotic Dynamic Systems using an Intelligent Tutoring System based on Fuzzy Logic and Fractal Theory", Proceedings of AIENG'97, Wessex Institute of Technology, pp. 97-100.

Castillo, O. & Melin, P. (1997b). "Mathematical Modelling and Simulation of Robotic Dynamic Systems using Fuzzy Logic Techniques and Fractal Theory", Proceedings of IMACS World Congress'97, Wissenschaft & Technik Verlag, Vol. 5, pp.343-348.

Castillo, O. & Melin, P. (1998a). "A New Fuzzy-Fractal-Genetic Method for Automated Mathematical Modelling and Simulation of Robotic Dynamic Systems", Proceedings of World Congress on Computational Intelligence FUZZ'98, IEEE Computer Society Press, Vol. 2, pp. 1182-1187.

Castillo, O. & Melin, P. (1998b). "Modelling, Simulation and Behavior Identification of Non-Linear Dynamical Systems with a New Fuzzy-Fractal-Genetic Approach", Proceedings of IPMU'98, EDK Publishers, Vol. 1, pp. 467-474.

Castillo, O. & Melin, P. (1998c). "Modelling, Simulation and Forecasting of International Trade Dynamics using a New Fuzzy-Genetic Approach", Proceedings of CCM'98, AMSE Press, pp. 21-24.

Chen, V. C. & Pao, Y. H. (1989). "Learning Control with Neural Networks", Proceedings of the International Conference on Robotics and Automation, pp. 1448-1453.

Chiu, S., Chand, S., Moore, D. & Chaudhary, A. (1991). "Fuzzy Logic for Control of Roll and Moment for a Flexible Wing Aircraft", IEEE Control Systems Magazine, Vol. 11, pp. 42-48.

Cybenko, G. (1989). "Approximation by Superpositions of a Sigmoidal Function", Mathematics of Control, Signals and Systems, Vol. 2, pp. 303-314.

Davidor, Y. (1991). "Genetic Algorithms and Robotics: A Heuristic Strategy for Optimization", World Scientific Publishing.

Devaney, R. (1989). "An Introduction to Chaotic Dynamical Systems", Addison Wesley Publishing.

Fahlman, S. E. & Lebiere C. (1990). "The Cascade-Correlation Learning Architecture", Advances in Neural Information Processing Systems, Morgan Kaufmann.

Fu, K.S., Gonzalez, R.C. & Lee, C.S.G (1987). "Robotics: Control, Sensing, Vision and Intelligence", McGraw-Hill.

Geman, S. & Geman, D. (1984). "Stochastic Relaxation, Gibbs Distribution and the Bayesian Restoration in Images", IEEE Transactions of Pattern Analysis and Machine Intelligence, Vol. 6, pp. 721-741.

Goldberg, D.E. (1989). "Genetic Algorithms in Search, Optimization and Machine Learning", Addison Wesley Publishing.

Grebogi, C., Ott, E. & Yorke, J. A. (1987). "Chaos, Strange Attractors, and Fractal Basin Boundaries in Nonlinear Dynamics", Science, Vol. 238, pp. 632-637.

Gujarati, D. (1987). "Basic Econometrics", McGraw-Hill Publishing.

Gupta, M. M. & Sinha, N. K. (1996). "Intelligent Control Systems: Theory and Applications", IEEE Computer Society Press.

Hanselman D. & Littlefield B. (1995). "The Student Edition of MATLAB Version 4 : User's Guide", The Math-Works, Inc. Prentice-Hall.

Holland, J. H. (1975). "Adaptation in Natural and Artificial Systems", University of Michigan Press.

Hunt, K. J., Sbarbaro, D., Zbikowski R. & Gawthrop, P. J. (1992). "Neural Networks for Control Systems-A survey", Automatica, Vol. 28 No. 6, pp. 1083-1112.

Ingber, L. & Rosen, B.E. (1992). "Genetic Algorithms and Very Fast Simulated Reannealing", Journal of Mathematical and Computer Modelling, Vol. 16, pp. 87-100.

Jamshidi, M. (1997). "Large-Scale Systems: Modelling, Control and Fuzzy Logic", Prentice-Hall

Jang, J.-S. R. (1993). "ANFIS: Adaptive-Network-Based Fuzzy Inference Systems", IEEE Transactions on Systems, Man and Cybernetics", Vol. 23, pp. 665-685.

Jang, J.-S. R. & Gulley, N. (1997). "MATLAB: Fuzzy Logic Toolbox, User's Guide", The Math-Works, Inc.Publisher.

Jang, J.-S. R., Sun, C.-T. & Mizutani, E. (1997). "Neurofuzzy and Soft Computing: A Computational Approach to Learning and Machine Intelligence", Prentice-Hall.

Kandel, A. (1992). "Fuzzy Expert Systems", CRC Press Inc.

Kapitaniak, T. (1996). "Controlling Chaos: Theoretical and Practical Methods in Non-Linear Dynamics", Academic Press.

Kasai, Y. & Morimoto, Y (1988). "Electronically Controlled Continuously Variable Transmission", Proceedings of International Congress Transportation Electronics.

Kirkpatrick, S., Gelatt, C. D. & Vecchi, M. P. (1983). "Optimization by Simulated Annealing", Science, Vol. 220, pp. 671-680.

Korn, G. A. (1995). "Neural Networks and Fuzzy Logic Control on Personal Computers and Workstations", MIT Press.

Kosko, B. (1992). "Neural Networks and Fuzzy Systems: A Dynamical Systems Approach to Machine Intelligence", Prentice-Hall.

Kosko, B. (1997). "Fuzzy Engineering", Prentice-Hall.

Lilly, K. W. (1993). "Efficient Dynamic Simulation of Robotic Mechanisms", Kluwer Academic Press.

Lim, S.Y., Hu, J. & Dawson, D.M. (1996). "An Output Feedback Controller for Trajectory Tracking of RLED Robots using Observed Backstepping Approach", Journal of Robotics and Automation, pp. 149-160.

Lippmann, R. P. (1987). "An Introduction to Computing with Neural Networks", IEEE Acoustics, Speech, and Signal Processing Magazine, Vol. 4, pp. 4-22.

Mamdani, E. H. & Assilian, S. (1975). "An Experiment in Linguistic Synthesis with a Fuzzy Logic Controller", International Journal of Man-Machine Studies, Vol. 7, pp. 1-13.

Mandelbrot, B. (1987). "The Fractal Geometry of Nature", W. H. Freeman and Company.

Masters, T. (1993). "Practical Neural Network recipe in C++", Academic Press, Inc.

Melin, P. & Castillo, O. (1996). "Modelling and Simulation for Bacteria Growth Control in the Food Industry using Artificial Intelligence", Proceedings of CESA'96, Gerf EC Lille, pp. 676-681.

Melin, P. & Castillo, O. (1997a). " An Adaptive Model-Based Neural Network Controller for Biochemical Reactors in the Food Industry", Proceedings of Control'97, Acta Press, pp.147-150.

Melin, P. & Castillo, O. (1997b). "Mathematical Modelling and Simulation of Bacteria Growth in the Time and Space Domains using Artificial Intelligence, Dynamical Systems and Fractal Theory", Proceedings of AMS'97, Acta Press, pp. 484-487.

Melin, P. & Castillo, O. (1997c). "An Adaptive Neural Network System for Bacteria Growth Control in the Food Industry using Mathematical Modelling and Simulation", Proceedings of IMACS World Congress'97, Wissenschaft & Technik Verlag, Vol 4 pp. 203-208.

Melin, P. & Castillo, O. (1997d). "Automated Mathematical Modelling and Simulation for Bacteria Growth Control in the Food Industry using Artificial Intelligence and Fractal Theory", Journal of Systems Analysis, Modelling and Simulation, Edited by Achim Sydow, Gordon and Breach Publishers, Vol. 29, pp. 189-206.

Melin, P. & Castillo, O. (1998a). "An Adaptive Model-Based Neuro-Fuzzy-Fractal Controller for Biochemical Reactors in the Food Industry", Proceedings of IJCNN'98, IEEE Computer Society Press, Vol. 1, pp.106-111.

Melin, P. & Castillo, O. (1998b). "A New Method for Adaptive Model-Based Neuro-Fuzzy-Fractal Control of Non-Linear Dynamic Plants: The Case of Biochemical Reactors", Proceedings of IPMU'98, EDK Publishers, Vol. 1, pp. 475-482.

Melin, P. & Castillo, O. (1998c). "A New Method for Adaptive Model-Based Neuro-Fuzzy-Fractal Control of Non-Linear Dynamical Systems", Proceedings of the International Conference of Non-Linear Problems in Aviation and Aerospace'98, Daytona Beach, Florida, USA (to appear).

Miller, W. T., Sutton, R. S. & Werbos P. J. (1995). "Neural Networks for Control", MIT Press.

Minsky, M. & Papert, S. (1969). "Preceptrons", MIT Press.

Morari, M. & Zafiriou, E. (1989). "Robust Process Control", Prentice-Hall.

Moulet, M. (1992). "A Symbolic Algorithm for Computing Coefficients Accuracy in Regression", Proceedings of the International Workshop on Machine Learning, pp. 332-337.

Nakamura, S. (1997). "Numerical Analysis and Graphic Visualization with MATLAB", Prentice Hall.

Narendra, K. S. & Annaswamy, A. M. (1989). "Stable Adaptive Systems", Prentice Hall Publishing Company.

Ng, G. W. (1997). "Application of Neural Networks to Adaptive Control of Non-Linear Systems", John Wiley & Sons.

Omidvar, O. & Elliot, D. L. (1997). "Neural Systems for Control", Academic Press.

Otten, R. H. J. M. & van Ginneken, L. P. P. P. (1989). "The Annealing Algorithm", Kluwer Academic.

Pelczar, M. J. & Reid, R. D. (1982). "Microbiology", McGraw Hill.

Parker D. B. (1982). "Learning Logic", Invention Report S81-64, File 1, Office of Technology Licencing.

Pham, D. T. & Xing, L. (1995). "Neural Networks for Identification, Prediction and Control", Springer-Verlag.

Pomerleau, D. A. (1991). "Efficient Training of Artificial Neural Networks for Autonomous Navigation", Journal of Neural Computation, Vol. 3, pp. 88-97.

Psaltis, D., Sideris, A. & Yamamura, A. (1988). "A Multilayered Neural Network Controller", IEEE Control Systems Magazine, Vol. 8, pp.17-21.

Rao, R. B. & Lu, S. (1993). "A Knowledge-Based Equation Discovery System for Engineering Domains", IEEE Expert, pp. 37-42.

Rasband, S.N. (1990). "Chaotic Dynamics of Non-Linear Systems", Wiley Interscience.

Rich, E. & Knight, K. (1991). "Artificial Intelligence", McGraw-Hill.

Rosenblatt, F. (1962). "Principles of Neurodynamics: Perceptrons and the Theory of Brain Mechanisms", Spartan.

Ruelle, D. (1990). "Deterministic Chaos: The Science and the Fiction", Proc. Roy. Soc. London, Vol. 427, pp. 241-248.

Rumelhart, D. E., Hinton, G. E. & Williams, R. J. (1986). "Learning Internal Representations by Error Propagation", Parallel Distributed Processing: Explorations in the Microestructure of Cognition, Vol. 1, Chapter 8, pp. 318-362, MIT Press.

Runkler, T. A. & Glesner, M. (1994). "Defuzzification and Ranking in the Context of Membership Value Semantics, Rule Modality, and Measurement Theory", Proceedings of European Congress on Fuzzy and Intelligent Technologies.

Russell, S. & Norvig, P. (1995). "Artificial Intelligence: A Modern Approach", Prentice-Hall.

Sackinger, E., Boser, B. E., Bromley, J., LeCun, Y. & Jackel, L. D. (1992). "Application of the Anna Neural Network Chip to High-Speed Character Recognition", IEEE Transactions on Neural Networks, Vol. 3, pp. 498-505.

Samad, T. & Foslien, W. (1994). "Neural Networks as Generic Nonlinear Controllers", Proceedings of the World Congress on Neural Networks, pp. 191-194.

Sejnowski, T. J. & Rosenberg, C. R. (1987). "Parallel Networks that Learn to Pronounce English Text", Journal of Complex Systems, Vol. 1, pp. 145-168.

Sleeman, D. & Edwards, P. (1992). "Proceedings of the International Workshop on Machine Learning", Morgan Kauffman Publishers.

Soucek, B. (1991). "Neural and Intelligent Systems Integration: Fifth and Sixth Generation Integrated Reasoning Information Systems", John Wiley and Sons.

Staib, W. E. (1993). "The Intelligent Arc Furance: Neural Networks Revolutionize Steelmaking", Proceedings of the World Congress on Neural Networks, pp. 466-469.

Staib, W. E. & Staib, R. B. (1992). "The Intelligent Arc Furnace Controller: A Neural Network Electrode Position Optimization System for the Electric Arc Furnace", Proceedings of the International Conference on Neural Networks, Vol. 3, pp. 1-9.

Su, H. T. & McAvoy, T. J. (1993). "Neural Model Predictive Models of Nonlinear Chemical Processes", Proceedings of the 8th International Symposium on Intelligent Control, pp. 358-363.

Su, H. T., McAvoy, T. J. & Werbos, P. J. (1992). "Long-term Predictions of Chemical Processes using Recurrent Neural Networks: A Parallel Training Approach", Industrial & Engineering Chemistry Research, Vol. 31, pp. 1338-1352.

Sueda, N. & Iwamasa, M. (1995). "A Pilot System for Plant Control using Model-Based Reasoning", IEEE Expert, Vol. 10, No. 4, pp.24-31.

Sugeno, M. & Kang, G. T. (1988). "Structure Identification of Fuzzy Model", Journal of Fuzzy Sets and Systems", Vol. 28, pp. 15-33.

Szu, H. & Hartley, R. (1987). "Fast Simulated Annealing", Physics Letters, Vol. 122, pp. 157-162.

Takagi, T. & Sugeno, M. (1985). "Fuzzy Identification of Systems and its Applications to Modeling and Control", IEEE Transactions on Systems, Man and Cybernetics, Vol. 15, pp. 116-132.

Troudet, T. (1991). "Towards Practical Design using Neural Computation", Proceedings of the International Conference on Neural Networks, Vol. 2, pp. 675-681.

Tsukamoto, Y. (1979). "An Approach to Fuzzy Reasoning Method", In Gupta, M. M., Ragade, R. K. and Yager, R. R., editors, Advanced in Fuzzy Set Theory and Applications, pp. 137-149, North-Holland.

Ungar, L. H. (1995). "A Bioreactor Benchmark for Adaptive Network-Based Process Control", Neural Networks for Control, MIT Press, pp. 387-402.

Von Altrock, C. (1995). "Fuzzy Logic & Neuro Fuzzy Applications Explained", Prentice Hall.

Weigend, A. & Gershenfeld, N.A. (1994). "Time Series Prediction: Forecasting the Future and Understanding the Past", Addison Wesley Publishing.

Werbos, P. J. (1991). "An Overview of Neural Networks for Control", IEEE Control Systems Magazine, Vol. 11, pp. 40-41.

Werbos, P. J. (1974). "Beyond Regression: New Tools for Prediction and Analysis in the Behavioral Sciences", Ph.D. Thesis, Harvard University.

Widrow, B. & Stearns, D. (1985). "Adaptive Signal Processing", Prentice-Hall.

Yager, R.R. & Filev, D.P. (1993). "SLIDE: A Simple Adaptive Defuzzification Method", IEEE Transactions on Fuzzy Systems, Vol. 1, pp. 69-78.

Yamamoto, Y. & Yun, X. (1997). "A Modular Approach to Dynamic Modelling of a Class of Mobil Manipulators", Journal of Robotics and Automation. pp. 41-48.

Yasunobu, S. & Miyamoto, S. (1985). "Automatic Train Operation by Predictive Fuzzy Control", Industrial Applications of Fuzzy Control, pp.1-18, North Holland.

Yen, J., Langar, R. & Zadeh, L. A. (1995). "Industrial Applications of Fuzzy Control and Intelligent Systems", IEEE Computer Society Press.

Zadeh, L.A. (1965). "Fuzzy Sets", Journal of Information and Control, Vol. 8, pp. 338-353.

Zadeh, L.A. (1971). "Similarity Relations and Fuzzy Ordering", Journal of Information Sciences", Vol. 3, pp. 177-206.

Zadeh, L.A. (1973). "Outline of a New Approach to the Analysis of Complex Systems and Decision Processes", IEEE Transactions on Systems, Man and Cybernetics, Vol. 3, pp. 28-44.

Zomaya, A. Y. (1992). "Modelling and Simulation of Robot Manipulators: A Parallel Processing Approach", World Scientific Publishing.

Appendix A

Prototype Intelligent Systems for Automated Mathematical Modelling

In this Appendix, we show two computer programs that can be considered prototype intelligent systems for automated mathematical modelling. First, we show a prototype intelligent system for automated modelling of general non-linear dynamical systems. Then, we show a prototype intelligent system for the domain of robotic dynamic systems. The implementation of both computer programs was done in ARITY© PROLOG interpreter Version 6.00.86 for MS-DOS.

A.1 Automated Mathematical Modelling of Dynamical Systems

We show in this section a computer program, in the PROLOG programming language, based on the fuzzy-fractal method for automated mathematical modelling described in Chapter 5 of this book. The prototype intelligent system for automated modelling uses as input the fractal dimension of the time series and the complexity of the problem (number of variables), and obtains as a result the "best" mathematical model of the dynamical system under consideration. The file

of this computer program can be found in the floppy disk accompanying this book
(the name of the file is Osmodel.ari).

```
/* Prototype intelligent system for automated mathematical modelling using Fuzzy Logic
   by Oscar Castillo and Patricia Melin (written in ARITY PROLOG)   */


automated_modelling :-
                write('input the fractal dimension:'),
                read( Fractal_dim),
                write('input the complexity of the problem:'),
                read( Dim),
                trend( Fractal_dim, Trend),
                time_series( Fractal_dim, Time_series),
                periodic_part( Fractal_dim, Periodic_part),
                model_selected( Fractal_dim, Dim, Trend, Periodic_part, Model),
                write('a candidate model is the:'), write(Model), nl,
                write('because the time series is:'), write(Time_series).


/* Module for Time Series Analysis */

trend( Fractal_dim, linear) :-
                Fractal_dim > 0.8,
                Fractal_dim < 1.2, !.

trend( Fractal_dim, non_linear) :-
                Fractal_dim >= 1.2,
                Fractal_dim < 1.5.

time_series( Fractal_dim, smooth) :-
                Fractal_dim > 0.8,
                Fractal_dim < 1.2, !.

time_series( Fractal_dim, cyclic) :-
                Fractal_dim >= 1.2,
                Fractal_dim < 1.5, !.

time_series( Fractal_dim, erratic) :-
                Fractal_dim >= 1.5,
                Fractal_dim < 1.8, !.
```

```
time_series( Fractal_dim, chaotic) :-
                Fractal_dim >= 1.8,
                Fractal_dim =< 2.0.

periodic_part( Fractal_dim, null) :-
                Fractal_dim > 0.8,
                Fractal_dim < 1.2, !.

periodic_part( Fractal_dim, simple) :-
                Fractal_dim >= 1.2,
                Fractal_dim < 1.4, !.

periodic_part( Fractal_dim, regular) :-
                Fractal_dim >= 1.4,
                Fractal_dim < 1.6, !.

periodic_part( Fractal_dim, difficult) :-
                Fractal_dim >= 1.6,
                Fractal_dim < 1.7, !.

periodic_part( Fractal_dim, very_difficult) :-
                Fractal_dim >= 1.7,
                Fractal_dim < 1.8, !.

periodic_part( Fractal_dim, chaotic) :-
                Fractal_dim >= 1.8,
                Fractal_dim =< 2.0.


/* Module for Selection of the Mathematical Models */

model_selected( Fractal_dim, one, linear, null, linear_regression) :-
                Fractal_dim > 0.8,
                Fractal_dim < 1.1, !.

model_selected( Fractal_dim, one, linear, null, logarithmic_regression) :-
                Fractal_dim >= 1.1,
                Fractal_dim < 1.2.

model_selected( Fractal_dim, one, non_linear, simple, logistic_differential_equation) :-
                triangular_logistic( Fractal_dim, Membership),
                Membership > 0.5.

model_selected( Fractal_dim, two, non_linear, simple,
lotka_volterra_differential_equation) :-
                triangular_volterra_simple( Fractal_dim, Membership),
                Membership > 0.5.
```

model_selected(Fractal_dim, three, non_linear, regular, lorenz_differential_equation) :
                triangular_lorenz(Fractal_dim, Membership),
                Membership > 0.5.

model_selected(Fractal_dim, one, non_linear, simple, logistic_difference_equation) :-
                triangular_logistic(Fractal_dim, Membership),
                Membership > 0.6.

model_selected(Fractal_dim, two, non_linear, regular,
lotka_volterra_difference_equation) :-
                triangular_volterra_regular(Fractal_dim, Membership),
                Membership > 0.6.

/* Triangular Membership Functions for the Fuzzy Sets */

triangular_logistic(Fractal_dim, 0) :-
                Fractal_dim < 1.2, !.

triangular_logistic(Fractal_dim, Membership) :-
                Fractal_dim >= 1.2,
                Fractal_dim < 1.3, !,
                Membership is (Fractal_dim - 1.2)/(0.1).

triangular_logistic(Fractal_dim, Membership) :-
                Fractal_dim >= 1.3,
                Fractal_dim < 1.4, !,
                Membership is (1.4 - Fractal_dim)/(0.1).

triangular_logistic(Fractal_dim, 0) :-
                Fractal_dim >= 1.4.

triangular_volterra_simple(Fractal_dim, 0) :-
                Fractal_dim < 1.2, !.

triangular_volterra_simple(Fractal_dim, Membership) :-
                Fractal_dim >= 1.2,
                Fractal_dim < 1.3, !,
                Membership is (Fractal_dim - 1.2)/(0.1).

triangular_volterra_simple(Fractal_dim, Membership) :-
                Fractal_dim >= 1.3,
                Fractal_dim < 1.4, !,
                Membership is (1.4 - Fractal_dim)/(0.1).

```
triangular_volterra_simple( Fractal_dim, 0) :-
        Fractal_dim >= 1.4.

triangular_volterra_regular( Fractal_dim, 0) :-
        Fractal_dim < 1.4, !.

triangular_volterra_regular( Fractal_dim, Membership) :-
        Fractal_dim >= 1.4,
        Fractal_dim < 1.5, !,
        Membership is (Fractal_dim - 1.4)/(0.1).

triangular_volterra_regular( Fractal_dim, Membership) :-
        Fractal_dim >= 1.5,
        Fractal_dim < 1.6, !,
        Membership is (1.6 - Fractal_dim)/(0.1).

triangular_volterra_regular( Fractal_dim, 0) :-
        Fractal_dim >= 1.6.

triangular_lorenz( Fractal_dim, 0) :-
        Fractal_dim < 1.4, !.

triangular_lorenz( Fractal_dim, Membership) :-
        Fractal_dim >= 1.4,
        Fractal_dim < 1.5, !,
        Membership is (Fractal_dim - 1.4)/(0.1).

triangular_lorenz( Fractal_dim, Membership) :-
        Fractal_dim >= 1.5,
        Fractal_dim < 1.6, !,
        Membership is (1.6 - Fractal_dim)/(0.1).

triangular_lorenz( Fractal_dim, 0) :-
        Fractal_dim >= 1.6.
```

A.2 Automated Mathematical Modelling of Robotic Dynamic Systems

We show in this section a computer program written in PROLOG based on the fuzzy-fractal method for automated modelling (described in Chapter 5) for the domain of robotic dynamic systems. The prototype intelligent system for automated modelling of robotic dynamic systems uses as input the fractal dimension of the time series and the number of links of the robotic system, and obtains as a result the "best" mathematical model of the robotic system under consideration. The file of this computer program can be found in the floppy disk accompanying this book. (The name of the file is Prmorob1.txt).

```
/* Prototype intelligent system for automated mathematical modelling of Robotic
Dynamic Systems using Soft Computing techniques
 by Oscar Castillo and Patricia Melin
(written in the ARITY PROLOG programming language) */


automated_modelling_robotic :-
                write('input the fractal dimension of the time series:'),
                read( Fractal_dim),
                write('input the number of links of the robotic system:'),
                read( Dim),
                trend( Fractal_dim, Trend),
                time_series( Fractal_dim, Time_series),
                periodic_part( Fractal_dim, Periodic_part),
                model_selected( Fractal_dim, Dim, Trend, Periodic_part, Model),
                write('a candidate model for the robotic system is the:'), nl, write(Model), nl,
                write('because the time series is:'), nl, write(Time_series).


/* Module for Time Series Analysis */

trend( Fractal_dim, linear) :-
                Fractal_dim > 0.8,
                Fractal_dim < 1.2, !.

trend( Fractal_dim, non_linear) :-
                Fractal_dim >= 1.2,
                Fractal_dim =< 2.0.

time_series( Fractal_dim, smooth) :-
                Fractal_dim > 0.8,
                Fractal_dim < 1.2, !.

time_series( Fractal_dim, cyclic) :-
                Fractal_dim >= 1.2,
                Fractal_dim < 1.5, !.
```

```
time_series( Fractal_dim, erratic) :-
                    Fractal_dim >= 1.5,
                    Fractal_dim < 1.8, !.

time_series( Fractal_dim, chaotic) :-
                    Fractal_dim >= 1.8,
                    Fractal_dim =< 2.0.

periodic_part( Fractal_dim, null) :-
                    Fractal_dim > 0.8,
                    Fractal_dim < 1.2, !.

periodic_part( Fractal_dim, simple) :-
                    Fractal_dim >= 1.2,
                    Fractal_dim < 1.4, !.

periodic_part( Fractal_dim, regular) :-
                    Fractal_dim >= 1.4,
                    Fractal_dim < 1.6, !.

periodic_part( Fractal_dim, difficult) :-
                    Fractal_dim >= 1.6,
                    Fractal_dim < 1.7, !.

periodic_part( Fractal_dim, very_difficult) :-
                    Fractal_dim >= 1.7,
                    Fractal_dim < 1.8, !.

periodic_part( Fractal_dim, chaotic) :-
                    Fractal_dim >= 1.8,
                    Fractal_dim =< 2.0.


/* Module for Selection of the Mathematical Models */

model_selected( Fractal_dim, one, linear, null, linear_oscillator) :-
                    Fractal_dim > 0.8,
                    Fractal_dim < 1.2.

model_selected( Fractal_dim, one, non_linear, simple, quadratic_oscillator) :-
                    triangular_quadratic( Fractal_dim, Membership).

model_selected( Fractal_dim, one, non_linear, regular, cubic_oscillator) :-
                    triangular_cubic( Fractal_dim, Membership).

model_selected( Fractal_dim, one, non_linear, difficult, forced_quadratic_oscillator) :-
                    triangular_forced_quad( Fractal_dim, Membership).

model_selected( Fractal_dim, one, non_linear, very_difficult, forced_cubic_oscillator) :-
                    triangular_forced_cub( Fractal_dim, Membership).
```

```prolog
model_selected( Fractal_dim, one, non_linear, chaotic, strongly_forced_oscillator) :-
            triangular_strongly_forced( Fractal_dim, Membership).

model_selected( Fractal_dim, two, linear, null, double_linear_oscillators) :-
            Fractal_dim > 0.8,
            Fractal_dim < 1.2.

model_selected( Fractal_dim, two, non_linear, simple, coupled_quadratic_oscillators) :-
            triangular_quadratic( Fractal_dim, Membership).

model_selected( Fractal_dim, two, non_linear, regular, coupled_cubic_oscillators) :-
            triangular_cubic( Fractal_dim, Membership).

model_selected( Fractal_dim, two, non_linear, difficult, coupled_forced_quad_oscillators) :-
            triangular_forced_quad( Fractal_dim, Membership).

model_selected( Fractal_dim, two, non_linear, very_difficult, coupled_forced_cub_oscillators) :-
            triangular_forced_cub( Fractal_dim, Membership).

model_selected( Fractal_dim, two, non_linear, chaotic, coupled_strongly_forced_oscillators) :-
            triangular_strongly_forced( Fractal_dim, Membership).

model_selected( Fractal_dim, three, linear, null, triple_linear_oscillators) :-
            Fractal_dim > 0.8,
            Fractal_dim < 1.2.

model_selected( Fractal_dim, three, non_linear, simple, three_coupled_quad_oscillators) :-
            triangular_quadratic( Fractal_dim, Membership).

model_selected( Fractal_dim, three, non_linear, regular, three_coupled_cubic_oscillators) :-
            triangular_cubic( Fractal_dim, Membership).

model_selected( Fractal_dim, three, non_linear, difficult, three_coupled_force_quad_oscillators) :-
            triangular_forced_quad( Fractal_dim, Membership).

model_selected( Fractal_dim, three, non_linear, very_difficult, three_coupled_force_cub_oscill) :-
            triangular_forced_cub( Fractal_dim, Membership).

model_selected( Fractal_dim, three, non_linear, chaotic, three_coupled_strongly_forced_oscill) :-
            triangular_strongly_forced( Fractal_dim, Membership).


/* Triangular Membership Functions for the Fuzzy Sets*/

triangular_quadratic( Fractal_dim, 0) :-
            Fractal_dim < 1.2, !.

triangular_quadratic( Fractal_dim, Membership) :-
            Fractal_dim >= 1.2,
            Fractal_dim < 1.3, !,
            Membership is ( Fractal_dim - 1.2)/(0.1).
```

```
triangular_quadratic( Fractal_dim, Membership) :-
                Fractal_dim >= 1.3,
                Fractal_dim < 1.4, !,
                Membership is ( 1.4 - Fractal_dim)/(0.1).

triangular_quadratic( Fractal_dim, 0) :-
                Fractal_dim >= 1.4.

triangular_cubic( Fractal_dim, 0) :-
                Fractal_dim < 1.4, !.

triangular_cubic( Fractal_dim, Membership) :-
                Fractal_dim >= 1.4,
                Fractal_dim < 1.5, !,
                Membership is ( Fractal_dim - 1.4)/(0.1).

triangular_cubic( Fractal_dim, Membership) :-
                Fractal_dim >= 1.5,
                Fractal_dim < 1.6, !,
                Membership is ( 1.6 - Fractal_dim)/(0.1).

triangular_cubic( Fractal_dim, 0) :-
                Fractal_dim >= 1.6.

triangular_forced_quad( Fractal_dim, 0) :-
                Fractal_dim < 1.6, !.

triangular_forced_quad( Fractal_dim, Membership) :-
                Fractal_dim >= 1.6,
                Fractal_dim < 1.65, !,
                Membership is ( Fractal_dim - 1.6)/(0.05).

triangular_forced_quad( Fractal_dim, Membership) :-
                Fractal_dim >= 1.65,
                Fractal_dim < 1.7, !,
                Membership is ( 1.7 - Fractal_dim)/(0.05).

triangular_forced_quad( Fractal_dim, 0) :-
                Fractal_dim >= 1.7.

triangular_forced_cub( Fractal_dim, 0) :-
                Fractal_dim < 1.7, !.

triangular_forced_cub( Fractal_dim, Membership) :-
                Fractal_dim >= 1.7,
                Fractal_dim < 1.75, !,
                Membership is ( Fractal_dim - 1.7)/(0.05).
```

```
triangular_forced_cub( Fractal_dim, Membership) :-
                Fractal_dim >= 1.75,
                Fractal_dim < 1.8, !,
                Membership is ( 1.8 - Fractal_dim)/(0.05).

triangular_forced_cub( Fractal_dim, 0) :-
                Fractal_dim >= 1.8.

triangular_strongly_forced( Fractal_dim, 0) :-
                Fractal_dim < 1.8, !.

triangular_strongly_forced( Fractal_dim, Membership) :-
                Fractal_dim >= 1.8,
                Fractal_dim < 1.9, !,
                Membership is ( Fractal_dim - 1.8)/(0.1).

triangular_strongly_forced( Fractal_dim, Membership) :-
                Fractal_dim >= 1.9,
                Fractal_dim =< 2.0, !,
                Membership is ( 2.0 - Fractal_dim)/(0.1).

triangular_strongly_forced( Fractal_dim, 0) :-
                Fractal_dim > 2.0.
```

Appendix B

Prototype Intelligent Systems for Automated Simulation

In this appendix, we show a prototype intelligent system for automated simulation of non-linear dynamical systems written in the ARITY© PROLOG programming language. We also show computer programs written in the MATLAB programming language (Version 5.1 for Windows 95) for the simulation of several non-linear dynamical systems.

B.1 Automated Simulation of Non-Linear Dynamical Systems

We show in this section a computer written in PROLOG based on our method for automated simulation using our new fuzzy-genetic approach (described in Chapter 6 of this book). The computer program uses as input the initial population and the maximum number of iterations (of the genetic algorithm), and obtains as a result the final population of parameters as well as the corresponding behavior identification for the dynamical system. The file of this computer program can be found in the floppy disk accompanying this book under the name of Osimrob3.txt.

```
/* Prototype Intelligent System for Automated Simulation of Dynamical Systems
by Oscar Castillo and Patricia Melin
(developed in ARITY PROLOG)  */


automated_simulation :-
                write( ' Input the initial population:'),
                read( Initial_Population),
                write( 'Input the maximum number of iterations:'),
                read( NumberIter),
                parameter_selection( Initial_Population,NumberIter,Final_Population),
                behavior_identification( Final_Population).


/* parameter selection using a specific genetic algorithm */

parameter_selection( Initial_Population,NumberIter,Final_Population) :-
                NumberIter > 0,
                fitness_value( Initial_Population,List_fitness),
                select_two( Initial_Population,List_fitness,X,Y),
                crossover( Initial_Population,X,Y,Next_Population),
                mutation( Next_Population,Mutated_Population),
                NumberIter1 is NumberIter - 1,
                parameter_selection( Mutated_Population,NumberIter1,Final_Population).

parameter_selection( Final_Population,_,Final_Population).

fitness_value( [ ],[ ]).
fitness_value( [X | Initial_Population],[FitnessX | List]) :-
                evaluate(X,FitnessX),
                fitness_value( Initial_Population,List).

select_two( [X,Y | _],List_fitness,X,Y).
crossover( [_,_ | Rest],X,Y,[NewX,NewY | Rest]) :-
                split( X,X1,X2),
                split( Y,Y1,Y2),
                conc( X1,Y2,NewX),
                conc( Y1,X2,NewY).

mutation( [X | Rest],[NewX | Rest]) :-
                X = [ X1 | L],
                X1 = 0, !,
                NewX = [ 1 | L].
mutation( [X | Rest],[NewX | Rest]) :-
                X = [ X1 | L],
                X1 = 1,
                NewX = [ 0 | L].
```

```
split( List,L1,L2) :-
                member( M1,List),
                member( M2,List),
                L1 = [ M1,M2],
                conc( L1,L2,List).
```

/* evaluation of the fitness of the members of the population */

```
evaluate( X,FitnessX) :-
                dictionary( X,NewX),
                model( NewX,Iterate),
                classification( Iterate,Identification),
                fitness( Identification,FitnessX).
```

```
conc( [ ],L,L).
conc( [X | L1],L2,[X | L3]) :-
                conc(L1,L2,L3).
```

```
member( X,[X | L]).
member( X,[Y | L]) :-
                member( X,L).
```


/* dynamic behavior identification using a fuzzy rule base */

```
behavior_identification( [ ]).
behavior_identification( [X | Final_Population]) :-
                dictionary( X,NewX),
                numerical_solution( NewX,Identification),
                write(' The identification for the value:'),
                write( NewX), nl,
                write(' is the behavior known as:'),
                write(Identification), nl,
                behavior_identification( Final_Population).
```

```
numerical_solution( NewX,Identification) :-
                model( NewX,Iterate),
                classification( Iterate,Identification).
```


/* dictionary for decodification */

```
dictionary( [0,0,0,0],0).
dictionary( [0,0,0,1],1).
dictionary( [0,0,1,0],2).
dictionary( [0,0,1,1],3).
dictionary( [0,1,0,0],4).
dictionary( [0,1,0,1],5).
dictionary( [0,1,1,0],6).
dictionary( [0,1,1,1],7).
```

```
dictionary( [1,0,0,0],8).
dictionary( [1,0,0,1],9).
dictionary( [1,0,1,0],10).
dictionary( [1,0,1,1],11).
dictionary( [1,1,0,0],12).
dictionary( [1,1,0,1],13).
dictionary( [1,1,1,0],14).
dictionary( [1,1,1,1],15).


/*  particular mathematical model  */

model( 0,0).
model( 1,0).
model( 2,1).
model( 3,1).
model( 4,2).
model( 5,2).
model( 6,3).
model( 7,3).
model( 8,4).
model( 9,4).
model( 10,4).
model( 11,4).
model( 12,4).
model( 13,3).
model( 14,3).
model( 15,4).


/*  classification with a rule base  */

classification( 0,fixed_point).
classification( 1,cycle_period_two).
classification( 2,cycle_period_four).
classification( 3,cycle_period_eight).
classification( 4,chaotic_behavior).


/*  heuristic to evaluate the fitness  */

fitness( fixed_point,1).
fitness( cycle_period_two,2).
fitness( cycle_period_four,4).
fitness( cycle_period_eight,8).
fitness( chaotic_behavior,10).
```

B.2 Numerical Simulation of Non-Linear Dynamical Systems

We show in this section the computer programs, written in the MATLAB programming language, for the numerical simulation of biochemical reactors. These computer programs were used to obtain the simulation results shown in Section 8.2 of this book. The computer programs for the simulation of the other applications are contained in the floppy disk, but they aren't described here because they are very similar to the ones for biochemical reactors. In all cases, we use a Runge-Kutta method for the numerical solution of the differential equations. The only difference between the files for the different applications is in the mathematical models that are used. For the case of biochemical reactors we show below the computer programs for the simulation of the mathematical models (the names of these files can be found in the list.txt file in the floppy disk).

```
% Simulation of Model M1: one Bacteria used for Food Production.
% see Figure 8.13 of the Book.
% Modelling, Simulation and Control of Non-Linear Dynamical Systems.
% Patricia Melin and Oscar Castillo, 1998.
[t,y] = ode45('Modelsel3',[0 10],[97.5; 0]);
plot(t,y(:,1));
title('Simulation of M1: one Bacteria used for food production');
xlabel('time t (seconds)');
ylabel('Population of Bacteria N')
zoom
```

```
% Simulation of Model M1: one Bacteria used for Food Production.
% see Figure 8.14 of the Book.
% Modelling, Simulation and Control of Non-Linear Dynamical Systems.
% Patricia Melin and Oscar Castillo, 1998.
[t,y] = ode45('Modelsel3',[0 10],[97.5; 0]);
plot(t,y(:,2),'LineWidth',2);
title('Simulation of M1: one Bacteria used for food production');
xlabel('time t (seconds)');
ylabel('Product P')
zoom
```

```
function dy = Modelsel3(t,y)
dy = [30*y(1) - 0.3*y(1)^2 - 0.8*y(1); 0.8*y(1)];
```

```
% Simulation of Model M2: two Bacteria used for Food Production.
% see Figure 8.15 of the Book.
% Modelling, Simulation and Control of Non-Linear Dynamical Systems.
% Patricia Melin and Oscar Castillo, 1998.
[t,y] = ode45('Moselc2',[0 15],[26.5; 26.5; 0]);
plot(t,y(:,1));
title('Simulation of M2: two Bacteria used for food production');
xlabel('time t');
ylabel('Population of Bacteria N1')
zoom
```

```
function dy = Moselc2(t,y)
dy = [30*y(1)-0.3*y(1)^2-0.8*y(2)*y(1)-0.8*y(1);
    30*y(2)-0.3*y(2)^2-0.8*y(1)*y(2)-0.8*y(2);
    0.8*y(1)+0.8*y(2)];
```

```
% Simulation of Model M3: two good Bacteria (N1, N2) used for Food Production .
% and one "bad" Bacteria (N3).
% see Figure 8.16 of the Book.
% Modelling, Simulation and Control of Non-Linear Dynamical Systems.
% Patricia Melin and Oscar Castillo, 1998.
[t,y] = ode45('Mosel',[0 8],[65; 6.5; 10; 0]);
plot(t,y(:,1),'o',t,y(:,2),'+',t,y(:,3),'-');
title('Simulation of M3: two "good" Bacteria and one "bad" Bacteria');
xlabel('time t');
ylabel('Population of Bacteria, oo N1, ++ N2, -- N3')
```

```
function dy = Mosel(t,y)
dy = [30*y(1)-0.3*y(1)^2-0.8*y(2)*y(1)-0.8*y(1)-0.2*y(1)*y(3);
    30*y(2)-0.3*y(2)^2-0.8*y(1)*y(2)-0.8*y(2)-0.2*y(2)*y(3);
    30*y(3)-0.3*y(3)^2+0.2*y(1)*y(3)+0.2*y(2)*y(3);
    0.8*y(1)+0.8*y(2)];
```

% Simulation of Model M3: two good Bacteria (N1, N2) used for Food Production .
% and one "bad" Bacteria (N3) Case b.
% see Figure 8.17 of the Book.
% Modelling, Simulation and Control of Non-Linear Dynamical Systems.
% Patricia Melin and Oscar Castillo, 1998.
[t,y] = ode45('Mosel1',[0 2],[60; 60; 0.5; 0]);
plot(t,y(:,1),'o',t,y(:,2),'+',t,y(:,3),'-');
title('Simulation of M3: two "good" Bacteria and one "bad" Bacteria');
xlabel('time t');
ylabel('Population of Bacteria, oo N1, ++ N2, -- N3')

function dy = Mosel1(t,y)
dy = [50*y(1)-0.3*y(1)^2-0.8*y(2)*y(1)-0.8*y(1)-0.0002*y(1)*y(3);
    60*y(2)-0.3*y(2)^2-0.8*y(1)*y(2)-0.8*y(2)-0.0002*y(2)*y(3);
    20*y(3)-0.3*y(3)^2+0.0002*y(1)*y(3)+0.0002*y(2)*y(3);
    0.8*y(1)+0.8*y(2)];

Appendix C

Prototype Intelligent Systems for Adaptive Model-Based Control

In this Appendix, we show computer programs for adaptive model-based control of non-linear dynamical systems. First, we show a computer program for fuzzy logic model selection that was developed using the Fuzzy Logic Toolbox™ of the MATLAB© programming language. Then, we show computer programs for identification and control using neural networks that were developed with the Neural Networks Toolbox™ of the MATLAB programming language.

C.1 Fuzzy Logic Model Selection

We show in this section a computer program in MATLAB based on our new method for fuzzy model selection (described in Chapter 7 of this book) for the domain of biochemical reactors. The computer programs for model selection for the other applications can be obtained in a similar way. The file of this computer program can be found in the floppy disk accompanying this book, under the name modelsel.fis.

%Computer Program for Fuzzy Logic Model Selection for Adaptive Model-Based Control of biochemical reactors.

% Written by: Patricia Melin and Oscar Castillo in MATLAB© Version 5.1.

% See Figures 9.7-9.10 of the book.

% Program for model selection using the Fuzzy Inference System (FIS) from Fuzzy Logic Toolbox™.

```
[System]
Name='modelsel'
Type='mamdani'
NumInputs=1
NumOutputs=1
NumRules=3
AndMethod='min'
OrMethod='max'
ImpMethod='min'
AggMethod='max'
DefuzzMethod='centroid'

[Input1]
Name='temperature'
Range=[100 120]
NumMFs=3
MF1='Low':'gaussmf',[4.247 100]
MF2='Medium':'gaussmf',[4.247 110]
MF3='High':'gaussmf',[4.247 120]
```

[Output1]

Name='model'

Range=[100 120]

NumMFs=3

MF1='M1':'gaussmf',[4.247 100]

MF2='M2':'gaussmf',[4.247 110]

MF3='M3':'gaussmf',[4.247 120]

[Rules]

1, 1 (1) : 1

2, 2 (1) : 1

3, 3 (1) : 1


```
% membership functions
function y = gauss_mf(x, parameter)
%GAUSS_MF Gaussian membership function with two parameters.
%   GAUSSIAN(x, [sigma, c]) returns a matrix y with the same size
%   as x; each element of y is a grade of membership.

c = parameter(1);
sigma = parameter(2);
tmp = (x - c)/sigma;
y = exp(-tmp.*tmp/2);

function y = sig_mf(x, parameter)
%SIG_MF Sigmoidal membership function with two parameters.
%   SIG_MF(x, [a, c]) returns a matrix y with the same size
%   as x; each element of y is a grade of membership.

a = parameter(1); c = parameter(2);
y = 1./(1 + exp(-a*(x-c)));
```

C.2 Neural Networks for Identification and Control

We show in this section, computer programs for identification and control using neural networks that were developed with the Neural Networks Toolbox™ of MATLAB. In both cases, the backpropagation learning algorithm was used for training the neural networks with real data. We will only show the computer programs for the case of biochemical reactors. These computer programs were used to obtain the simulation results shown in Section 9.2 of this book. The computer programs for the other applications can be found in the floppy disk accompanying this book (the name of these files can be found in the list.txt file in the floppy disk). The names of the files for the case of biochemical reactors are traida1.m, traida3.m and traicon1.m.

```
% Computer Program for training the Neural Network for identification.
% See Figures 9.13, 9.14 and 9.15 of the Book.
% Modellig, Simulation and Control of Non-Linear Dynamical Systems.
% Patricia Melin and Oscar Castillo, 1998.
P = [ 0.0001 0.0003 0.0926 0.1170 0.1415 0.1905 0.2238  0.2570 0.2903 0.3236 0.3532...
      0.3827 0.4123 0.4418 0.4714 0.5009 0.5305 0.5893 0.6186 0.6479 0.6772 0.7049...
      0.7326 0.7880 0.8147 0.8414 0.8947 0.9222 0.9497 0.9772 1.0047 1.0340 1.0926...
      1.1219 1.1518 1.1818 1.2117 1.2416 1.2986 1.3270 1.3555 1.3822 1.4088 1.4355...
      1.4888 1.5155 1.5422 1.5974 1.6260 1.6546 1.6832 1.7135 1.7438 1.7740 1.8043...
      1.8338 1.8928 1.9223 2.0000];
T = [ 26.5002 26.5002 26.5403 26.5408 26.5447 26.5434 26.5492 26.5510 26.5439...
      26.5398 26.5491 26.5524 26.5436 26.5380 26.5503 26.5557 26.5430 26.5355...
      26.5514 26.5571 26.5424 26.5329 26.5550 26.5426 26.5342 26.5517 26.5433...
      26.5375 26.5473 26.5511 26.5437 26.5494 26.5533 26.5433 26.5369 26.5518...
      26.5570 26.5332 26.5506 26.5571 26.5422 26.5324 26.5469 26.5429 26.5362...
      26.5464 26.5437 26.5437 26.5389 26.5483 26.5518 26.5437 26.5384 26.5513...
      26.5558 26.5427 26.5525 26.5589 26.5425];
[W1,b1,W2,b2] = initff(P,5,'tansig',T,'tansig');
A2 = simuff(P,W1,b1,'tansig',W2,b2,'tansig');
disp_freq=5000
max_epoch=40000;
err_goal=0.00002;
lr=0.00001;
tp = [disp_freq max_epoch err_goal lr];
[W1,b1,W2,b2,epochs,tr] = trainbp(W1,b1,'tansig',W2,b2,'purelin',P,T,tp);
```

```
% Computer Program for training the Neural Network for identification.
% See Figures 9.16 and 9.17 of the Book.
% Modelling, Simulation and Control of Non-Linear Dynamical Systems.
% Patricia Melin and Oscar Castillo, 1998.
P = [0.0001 0.0003 0.0926 0.1170 0.1415 0.1905 0.2238  0.2570 0.2903 0.3236 0.3532..
    0.3827 0.4123 0.4418 0.4714 0.5009 0.5305 0.5893 0.6186 0.6479 0.6772 0.7049...
    0.7326 0.7880 0.8147 0.8414 0.8947 0.9222 0.9497 0.9772 1.0047 1.0340 1.0926...
    1.1219 1.1518 1.1818 1.2117 1.2416 1.2986 1.3270 1.3555 1.3822 1.4088 1.4355...
    1.4888 1.5155 1.5422 1.5974 1.6260 1.6546 1.6832 1.7135 1.7438 1.7740 1.8043...
    1.8338 1.8928 1.9223 2.0000];
T = [ 26.5002 26.5002 26.5403 26.5408 26.5447 26.5434 26.5492 26.5510 26.5439...
    26.5398 26.5491 26.5524 26.5436 26.5380 26.5503 26.5557 26.5430 26.5355...
    26.5514 26.5571 26.5424 26.5329 26.5550 26.5426 26.5342 26.5517 26.5433...
    26.5375 26.5473 26.5511 26.5437 26.5494 26.5533 26.5433 26.5369 26.5518....
    26.5570 26.5332 26.5506 26.5571 26.5422 26.5324 26.5469 26.5429 26.5362...
    26.5464 26.5437 26.5437 26.5389 26.5483 26.5518 26.5437 26.5384 26.5513...
    26.5558 26.5427 26.5525 26.5589 26.5425];
[W1,b1,W2,b2] = initff(P,5,'tansig',T,'tansig');
A2 = simuff(P,W1,b1,'tansig',W2,b2,'tansig');
disp_freq=5000
max_epoch=80000;
err_goal=0.00002;
lr=0.001;
tp = [disp_freq max_epoch err_goal lr];
[W1,b1,W2,b2,epochs,tr] = trainbp(W1,b1,'tansig',W2,b2,'purelin',P,T,tp)

% Computer Program for training the Neural Network for Control.
% See Figures 9.18 and 9.19 of the Book.
% Modelling, Simulation and Control of Non-Linear Dynamical Systems.
% Patricia Melin and Oscar Castillo, 1998.
P = [0.0001 0.1170 0.2238 0.3236 0.4123 0.5009 0.6186 0.7049 0.8147 0.9222 1.0047..
    1.1219 1.2117 1.3270 1.4088 1.5155 1.6260 1.7135 1.8043 1.9223 2.0000];
T = [0.005 3.9293 8.0875 12.3289 16.2527 20.0185 23.7850 27.5170 32.2898 36.8664..
    40.3349 45.1585 48.9182 53.9417 57.5704 62.1006 66.6318 70.2739 74.0590...
    79.1370 83.2939];
[W1,b1,W2,b2] = initff(P,5,'purelin',T,'purelin');
A2 = simuff(P,W1,b1,'purelin',W2,b2,'purelin');
disp_freq=1000
max_epoch=5000;
err_goal=0.00002;
lr=0.0001;
tp = [disp_freq max_epoch err_goal lr];
[W1,b1,W2,b2,epochs,tr] = trainbp(W1,b1,'purelin',W2,b2,'purelin',P,T,tp);
```

Index

Printed and bound by CPI Group (UK) Ltd, Croydon, CR0 4YY

23/10/2024

01778249-0008